T0295224

HANDBOOK OF
MATHEMATICAL SCIENCE
COMMUNICATION

World Scientific Series on Science Communication

Print ISSN: 2737-4912
Online ISSN: 2737-4793

Editor-in-Chief
Hans Peter Peters
Forschungszentrum Jülich, Germany
h.p.peters@fz-juelich.de

Aims and Scope

The books published in this series deal with the public communication of science, i.e. with communication of and about science involving non-scientists and taking place in the public sphere. Possible topics include public communication and discourses about scientific knowledge, scientific projects or research fields, and communication about science as a social system and its interdependencies with the larger society.

This book series is open to analyses of all forms of public communication and interaction: journalism, public relations of science, blogs, social media, video-sharing platforms, science museums, public events, engagement activities, public deliberation and participation, citizen science and other collaborations between scientists and citizens, for example. Books may focus on content and processes of messages and discourses, on actors and their strategies, on channel and arena characteristics, on the reception, effects and use of public expertise, on public controversies, on inclusion of citizens in public discourses, and on issues of quality, ethics, and trust.

Typically, authors/editors will come from the academic field and have an academic audience in mind. But some of the books may also be relevant for communication professionals, scientists (as communicators), science managers and knowledge users, for example. Books may be based on specific research projects, deal with a relevant subjects by means of review of existing studies and theoretical discussion, or publish contributions of a relevant conference (proceedings).

Published

World Scientific Series on Science Communication – Volume 3

HANDBOOK OF MATHEMATICAL SCIENCE COMMUNICATION

Editors

Anna Maria Hartkopf
Erin Henning

Freie Universität Berlin, Germany

World Scientific

NEW JERSEY • LONDON • SINGAPORE • BEIJING • SHANGHAI • HONG KONG • TAIPEI • CHENNAI • TOKYO

Published by

World Scientific Publishing Co. Pte. Ltd.

5 Toh Tuck Link, Singapore 596224

USA office: 27 Warren Street, Suite 401-402, Hackensack, NJ 07601

UK office: 57 Shelton Street, Covent Garden, London WC2H 9HE

Library of Congress Cataloging-in-Publication Data

Names: Hartkopf, Anna Maria, editor. | Henning, Erin, editor.

Title: Handbook of mathematical science communication / editors,
    Anna Maria Hartkopf, Erin Henning.

Description: Hackensack, New Jersey : World Scientific, [2023] |
    Series: World Scientific series on science communication, 2737-4912 ; volume 3 |
    Includes bibliographical references and index.

Identifiers: LCCN 2022020535 | ISBN 9789811253065 (hardcover) |
    ISBN 9789811253072 (ebook for institutions) | ISBN 9789811253089 (ebook for individuals)

Subjects: LCSH: Communication in mathematics.

Classification: LCC QA41.4 .H36 2023 | DDC 306.4/5--dc23/eng20220723

LC record available at https://lccn.loc.gov/2022020535

British Library Cataloguing-in-Publication Data

A catalogue record for this book is available from the British Library.

For any available supplementary material, please visit
https://www.worldscientific.com/worldscibooks/10.1142/12747#t=suppl

Typeset by Stallion Press
Email: enquiries@stallionpress.com

Contents

Foreword

Could there be a better time for a book on communicating mathematics? Hardly, because the topic is one of the utmost importance at the moment. Every morning during the COVID-19 pandemic we faced discussions about exponential growth and statistical analysis of the current situations in hospitals and with regards to infection and death rates. But mathematics plays a key role in our daily lives and the progression of our society even outside the pandemic. Supercomputers are calculating possible climate scenarios, AI will soon be driving our cars and robots are likely to become the future of both medicine and caregiving. Mathematics is everywhere and thus, it has become more crucial than ever to understand its implications as well as the basic principles behind it.

To enhance this, it is important to involve all members of society, especially with a special focus on women as one can see a lack of female representation in mathematics-related fields of work, rather than just address schools and students. While it is well worth the effort to increase extracurricular activities and to emphasise improving the way math is taught at school, those efforts will fall short if the overall perspective of mathematics within society does not change.

But there is hope and this handbook is part of that hope — at least when we are looking at the efforts made when it comes to communicating mathematics. If we are looking at existing projects and formats on communicating mathematical sciences, we are seeing more of them taking the broader perspective into account. Some of those state-of-the-art communication approaches are presented in this book.

The projects depicted in this part of the handbook have in common that they search for an eye-level and bidirectional dialogue with the target groups rather than monodirectional knowledge dissemination. Furthermore, most of the formats described offer perspectives for participation and are trying to be as inclusive as possible. This is good news, because as in most fields, knowledge dissemination is not conducted in a top-down manner but rather focused on inspiring interest and inviting mathematical action among various target groups. Because math is for everyone.

Besides presenting best practice examples and discussing the right approaches to communicating about the field, the main goal of this book is to foster discourse in the field of Mathematical Science Communication. A dialogue is necessary to establish Mathematical Science Communication as a rigorous and respectable scientific discipline that is located at the interface of mathematics, its didactics and communication science. To achieve that, it is crucial to foster the exchange between the practice of science communication and the relatively new field of science communication research. The handbook opens up bidirectional dialogue about this topic and draws on similarities to another field at the intersection of mathematics and communication: didactics. As a field, Mathematical Science Communication can learn a lot when it comes to establishing itself as well as defining its twofold objectives: one is setting the agenda for research and the other is educating a new generation of mathematical science communicators.

Of course, the *Handbook of Mathematical Science Communication* can only be a starting point for this. Especially because the science of science communication is a relatively young research field and the need for it to further divide into differing scientific subjects is quite novel. But Erin Henning and Anna Maria Hartkopf deliver just that: an important collection of practices and research in the field of Mathematical Science Communication as well as a starting point for future discussions.

Markus Weißkopf
Managing Director
Wissenschaft im Dialog
Germany

Acknowledgment

We, the editors, would like to express our special thanks of gratitude to Hans Peter Peters for pointing us to the opportunity to publish this book. Furthermore, we are dearly obliged to Günter M. Ziegler for his support.

Lastly, we would like to thank the Collaborative Research Center "Discretization in Geometry and Dynamics" (CRC DGD), under the leadership of Alexander I. Bobenko and Klaus Tschira Stiftung, for backing our affiliation with Freie Universität Berlin.

https://doi.org/10.1142/9789811253072_0001

Introduction

Anna Maria Hartkopf* and Erin Henning†

*Institute of Mathematics, Freie Universität Berlin,
Berlin, Germany*
*anna.hartkopf@fu-berlin.de
†erin.henning@fu-berlin.de

The mathematics exhibition in Munich 1893 on the occasion of the Annual Meeting of the German Mathematical Society (founded in 1891), curated by Walther von Dyck, is one of the visible starting points of Mathematical Science Communication: it drew a large crowd to the Technische Hochschule München, where objects and results of mathematical research were displayed in four rooms for an interested public. One of the rooms presented a "Wunderkammer" of geometric models — they attracted the interest of the audience due to their aesthetically pleasing nature and the unusual, novel shapes they represented [1].

The start of the Public Understanding of Science and Humanities (PUSH) movement in the UK nearly four decades ago, as documented by the 1985 "Bodmer report", led to more concerted efforts in the mathematical scientific community to reach out to the public and let them in on the topics of their research. We can see an increase of projects that center around mathematics and aim to foster interest in these topics for people outside of the scientific realm. These projects come in various shapes and sizes: from exhibitions and entire museums, mathematical artwork, and films to projects that are more closely related to education, like workshops and rallies.

The aim of this book is twofold. On the one hand, we present a variety of recent projects that use reseach-backed approaches for the communication of mathematical science to various target groups. The projects have in common that they search for an eye-level dialogue with their participants and emphasize active participation. The dissemination of knowledge is not conducted in a top-down manner but rather focused on inspiring interest and inviting mathematical action. The other and more fundamental goal of this handbook is to foster discourse in the field of Mathematical Science Communication. This will help to promote exchange among the actors in the practice of science communication and the relatively new field of science communication research as well as to ensure that Mathematical Science Communication can itself develop into a rigorous and respectable scientific discipline that is located in the field between mathematics, its didactics and communication science. Similar to the establishment of mathematics didactics, the manifestation of Mathematical Science Communication as a scientific discipline needs to establish systematic terminology along with empirical methods to measure and compare the practical side. This, in turn, will allow the various actors behind the projects to systematically and rigorously develop future projects based on current research findings. This handbook can only be a starting point for this, as the field of the science of science communication is relatively young and the need to further divide it into the differing scientific subjects is quite novel.

When we look at the different actors behind many mathematical outreach projects, we can see that universities and research institutions are the major players behind the organization and conceptualization of outreach projects, especially during their founding periods. For many of these projects, most of the funding comes from government institutions and private foundations. However, some are implemented by other means, namely nonprofit organizations and motivated groups of private individuals consisting of, for example, mathematicians, teachers, and artists. Motivations are closely linked to values that undergird these endeavors. For example, an institute founding an exhibition is aiming to enhance its recognition in society or stand out in comparison to other institutions. Projects closer to formal education might be motivated by the wish to bring extracurricular learning opportunities to children that are not likely to experience them within their social surroundings and thus here we find equality as a guiding motif. By promoting certain competencies related to decision-making in the population, efforts can be driven by democratic values of general societal welfare.

Many of the projects in this handbook state "promoting the enjoyment of mathematics" and "raising awareness of the role mathematics plays in everyday life" as primary objectives. Others want to bridge the gap between research mathematics and formal education or to alter the image of mathematics held by the society in which the projects reside to reflect how mathematicians perceive mathematics. Some wanted to create physical or virtual spaces for their community to come together to experience and actively participate in mathematical thinking. Other objectives include advertising a particular institution or foundation, promoting competencies in society, highlighting women in mathematics, or filling a perceived gap in a certain area, such as a need for a mathematics museum in a particular region. Others aim to create content to complement school curricula or to enhance teacher training. The artists' intentions are to express how they perceive mathematics as well as bring life and a face to professional mathematicians in their projects.

For this handbook, we chose projects that attempt to engage their audience through an interactive, playful and engaging manner, many focusing on the visual side of mathematics to convey deeper mathematical concepts. They promote a "learn-by-doing" philosophy by leading participants to experiment and discover for themselves the underlying mathematics. Some utilize hands-on activities and creating to guide participants to these moments of realization while others connect mathematics to various fields, particularly art, to stimulate interest in the target group. Many make use of the internet as a platform to reach as wide an audience as possible and create content for teachers to be used in classrooms around the world. Above all, most of these projects emphasize an active audience and promote creating a dialogue amongst the participants and between society and mathematicians.

Most of the rather scarce literature about Mathematical Science Communication [2,3] implicitly rely on the so-called "deficit model" [4]: the belief that more knowledge will lead to greater acceptance and the assumption that experts' knowledge is always superior to other forms of knowledge [5]. These assumptions have been empirically disproven and now other formats are being developed that focus on building trust as a prerequisite for open dialogue.

In Mathematical Science Communication, we can observe that some assumptions are perpetuated for a long time without really being questioned. One is a strong emphasis on evoking an emotional response. There is talk about "beauty" and "love for mathematics" and a wish to share these positive feelings with others. Another term that is often mentioned is "mathematical myths". Under this term, the reasons for the ascribed negative attitudes toward the subject are summarized. This raises the question of who actually gets to decide on the "correct" perception of mathematics.

What is the future of Mathematical Science Communication? On the one hand, there undeniably is expert knowledge, and co-creation of knowledge should not discredit the scientific rigor of mathematics. On the other hand, we can see that certain forms of top-down communication are outdated or maybe suitable only for target groups that are already interested in learning. The task is to create innovative participatory and trust-building formats of Mathematical Science Communication that enable the transfer of knowledge and allow for dialogue at eye level.

One approach that has a long tradition in other sciences, especially in biodiversity research, is the concept of citizen science. Here participants take an active part in the research process, actively contribute to generating new knowledge, and answer open research questions. Especially in the collection of data, such as the mosquito atlas [6], or the analysis of large amounts of data, such as photographs of galaxies and the determination of their shapes [7], citizens can make a decisive contribution because the research tasks simply take too much time and require human creativity and decision skills. When applying the concept of citizen science to mathematics, the first of the many challenges one is faced with is to find a suitable problem that can be solved in many small steps by non-mathematically trained individuals [8].

Responsible journalism is important to a democratic society as it serves a watchdog function by providing citizens with information as well as holding institutions within the democracy accountable. Journalism is threatened as print is suffering from the shift to digital media and jobs are at risk in this area. However, presenting complex content to their audiences in an understandable and still entertaining manner produces a challenging task. Preselection, critical analysis, and unbiased commenting is the desired goal of quality journalism.

At the time this handbook was edited, the COVID-19 crisis was dominating the headlines and its impact on the field of science communication is yet to be fully determined. However, as many of our authors have mentioned in their chapters, the pandemic has dramatically altered how science communicators approach their target groups and it has highlighted the need for responsible, trustworthy science journalism.

The first section of this handbook consists Mathematical Science Communication projects that are related to mathematics education and lie on the spectrum between formal and informal learning. In Chapter 1, Ems Lord, the director of the NRICH project based at the University of Cambridge, provides insight into the development of the collaborative mathematics outreach initiative. It pioneered as an online project in the late 1990s and has developed into an extremely successful and continuously growing platform that directly addresses pupils as well as their teachers. The Experience Workshop is a collaboration between Finnish and Hungarian mathematics educators and artists that is described in Chapter 2 by Kristóf Fenyvesi *et al.* The workshop comprises various activities related to arts and mathematics and can, besides disseminating mathematical knowledge, be a vehicle for cultural integration, learning languages, and bringing together diverse groups. In Chapter 3, Milena Damrau emphasizes the importance of the role that mathematics teachers play as science communicators, as most people's perceptions of mathematics are shaped in the classroom. Damrau argues that future mathematics teachers may benefit from attending science communication lectures during their studies to prepare them for this, in addition to their primary task of formally teaching mathematical competencies. One example of a Mathematical Science Communication project designed specifically for the use of teachers to complement formal education is described in Chapter 4. Francien Bossema *et al.* give a detailed account of the implementation of a math trail through the city of Leiden and provide a guideline to develop a similar endeavor.

Section 2 is comprised of a collection of detailed accounts on the founding, implementation, and day-to-day operations behind science museums and exhibitions dedicated to mathematics. It begins with the well-established National Museum of Mathematics, better known as the MoMath in New York City where co-founder Glen Whitney discusses the road taken to bring such a museum into a thriving existence in Chapter 5. In Chapter 6, Sylvie Benzoni and Marion Liewig present the current and ongoing efforts behind the new Maison Poincaré exhibition in Paris, supported by the Institut Henri Poincaré, set to open in 2023. The authors provide a detailed insight into the interplay of the design and curation process that leads to the meticulously planned display. Chapter 7 presents three perspectives involved in the founding of the *Mathematics Adventure Land* exhibition. First, Bernhard Ganter describes the process and motivation of the foundation of the museum, Rahel Brugger details the participatory activities of the exhibition and finally, Andrea Hoffkamp draws the links between the museum and the mathematics teacher training of Technische Universität Dresden. In Chapter 8, Eric Londaits *et al.* tell the story of the nonprofit organization IMAGINARY from being a traveling exhibition founded by the Mathematical Research Institute of Oberwolfach to becoming a global platform of open and interactive mathematics exhibitions. The authors go on to give an account of their most recent and ongoing projects, such as *La La Lab* which centers around music and mathematics. Martin Skrodzki presents the *Illustrating Mathematics* program by ICERM at Brown University in Chapter 9, which brought together mathematicians and artists to develop new formats of visualization in research mathematics. In order to draw public attention to this special program, an exhibition was organized.

How the arts can be a successful tool to communicate mathematics in various formats is highlighted in Section 3. Constanza Rojas-Molina, who is primarily a mathematician and has an interest in science communication, uses her talents as a graphic illustrator to visualize mathematics and promote women in math. In Chapter 10, Rojas-Molina presents the process behind utilizing social media to communicate mathematics to a wide audience through recurring Twitter drawing challenges. Chapter 11 would have been a good fit for Section 1 as it has strong elements of mathematics education but we decided to insert it here because the emphasis on art is what makes it unique. Melissa Silk and Annette Mauer, directors of the STEAMpop traveling exhibition in Australia, discuss how the arts (A) can be used to promote the traditional STEM fields by conducting unconventional hands-on and artistic workshops to illustrate various mathematical concepts. In Chapter 12, contemporary filmmaker Ekaterina Eremenko presents how she balances the entertainment value of films and the rigorous scientific themes in mathematics to produce successful films today. Eremenko takes the reader with her on a unique and personal journey of shooting and producing her aesthetically attractive pieces. In Chapter 13, Italian mathematician Michele Emmer, one of the first filmmakers to produce films about mathematics, gives a personal narrative of his various projects running since the 1970s.

Finally, the authors in Section 4 take one step back and assume a more analytical point of view on various topics related to Mathematical Science Communication. Ines Lein and Mirjam Jenny outline in Chapter 14 the research behind using infographics to promote risk literacy in society. The authors provide examples of how the Robert Koch Institute in Germany employs these infographics to make risk analysis in medical interventions more clear and transparent to the public. Felix G. Rebitschek goes a step further in Chapter 15 to discuss how risk literacy is intertwined with algorithmic literacy, which denotes an individual's ability to conduct oneself in algorithm-based information architecture in a way that promotes their objectives. Rebitschek outlines a roadmap for navigating the general competence framework in order to reach a definition of smart-world literacy. In Chapter 16, "Mathematical Science Communication as a Strategy for Democratizing Algorithmic Governance", Florian Eyert takes on a sociological perspective to discuss the importance for citizens and politicians to be part of a dialogic communication with mathematical experts, as more and more complex mathematical models are being used in political decision-making and may carry value judgment. Eyert argues that for reflexivity, it is crucial that the mathematical expert understands themself as having political agency. Andreas Loos is a data scientist with the German weekly online newspaper *ZEIT ONLINE* and gives the reader some insight into the ever-increasingly important relationship between mathematics and science journalism in Chapter 17. Loos analyzes how articles about mathematics are received by the public as well as how science journalists can responsibly apply mathematical models to create media especially if the research for an article is based on large amounts of data. Lastly, in Chapter 18, Thomas Vogt presents the development of science communication projects in Germany since the PUSH memorandum in 1999 with a focus on mathematics, which took off in 2008 when Germany announced the "Year of Mathematics". Vogt describes successful formats and trends in science communication with an emphasis on math and predicts what the future holds for Mathematical Science Communication.

References

[1] Fischer, G. (1986). *Mathematische Modelle (Mathematical Models)*, 2 Bände. Braunschweig/Wiesbaden: Vieweg.

[2] Behrends, E., Crato, N. and Rodrigues, J. F. (2012). *Raising Public Awareness of Mathematics*. Springer.

[3] Howson, A. G. and Kahane, J.-P. (1990). *The Popularization of Mathematics*. Cambridge: Cambridge University Press.

[4] Wynne, B. (1995). "Public understanding of science". *Handbook of Science and Technology Studies*, 1, 361–388.

[5] Hartkopf, A. M. (2020). *Mathematical Science Communication*. Dissertation, Freie Universität Berlin.

[6] Friedrich-Löffler-Institut (FLI) and Leibniz Centre for Agricultural Landscape Research (ZALF) e.V. Mückenatlas. https://mueckenatlas.com/about/ [Accessed on August 18, 2021].

[7] Zooniverse. https://www.zooniverse.org/projects/zookeeper/galaxy-zoo [Accessed on August 18, 2021].

[8] Hartkopf, A. M. (2019). "Developments towards Mathematical Citizen Science". *Forum Citizen Science*, 2019.

Chapter 1

NRICH*

Ems Lord

Director of NRICH, University of Cambridge, UK
ell35@cam.ac.uk

For over two decades, the NRICH team has been developing rich mathematical teaching and learning activities for students, families, and teachers. This trailblazing, research-informed project welcomed over 30 million pageviews to its online activities over the last 12 months (*Source*: Google Analytics). NRICH also works face-to-face with thousands of students and their teachers each year. This chapter explores the evolution of the NRICH project, from its early days as an online mathematics club for students to becoming an award-winning leader in mathematics education today.

At a time when developing effective problem-solving skills is becoming an increasingly crucial focus for jurisdictions across the globe, NRICH's rich mathematical activities are helping to ensure that future generations of students develop and embed the types of transferable skills that will enable them to adapt and thrive in their rapidly changing world.

Background

The ongoing success of NRICH is built on its clear vision for mathematics education. NRICH aims to:

- Enrich and enhance the experience of the mathematics curriculum for all learners.
- Let learners develop mathematical thinking and problem-solving skills.
- Offer challenging, inspiring, and engaging activities.
- Show rich mathematics in meaningful contexts.
- Work in partnership with teachers, schools, and other educational settings to share expertise.

---

* The NRICH website can be found at https://nrich.maths.org/.

NRICH aims to reach as wide an audience as possible with its inspiring mathematical ideas and activities. Its online approach can be accessed by teachers, students, and their families around the globe, widening access to mathematical outreach and helping to address issues from equality of access to high quality support.

The rapid growth of the internet has had a huge part to play in NRICH's development. When the project was launched in the late 1990s, computers were becoming much more commonplace in our homes and classrooms, yet the internet was very much in its infancy and its potential had yet to be fully understood. The trailblazing decision to make NRICH an online club brought it a much wider audience than would have previously been possible. Although NRICH is based in the UK, its internet-based approach ensures that the project is truly multinational; around half of NRICH users are teaching and learning outside the UK from countries as far afield as New Zealand and Australia. Today's NRICH team regularly receives solutions to its online problems from students living in dozens of different countries, solutions that it publishes on its website for everyone to read and reflect upon.

Sharing student solutions is a very special aspect of NRICH's work. Although the project is very well known for its online approach toward mathematical teaching and learning, being online does not imply that mathematics should necessarily be regarded as a solitary activity. The NRICH team is keen to promote mathematics as a collaborative opportunity for students. Mathematicians frequently enjoy sharing and solving problems together; exploring one another's solutions enables them to make connections between different areas of mathematics, extend their knowledge, and develop the flexibility to solve a wider range of problems. Solving problems can be deeply satisfying, even more so when solving them with others. NRICH maximizes the potential of the internet by offering students the opportunity to share their solutions and ideas. The team receives emails from students around the world who are encouraged to share solutions explaining their reasoning as well as their final answers. Some students share written accounts of their work, while others send photographs of their work, spreadsheets or voice files. This eclectic approach values different forms of mathematical communication and helps to increase understanding about the usefulness of learning a range of mathematical representations. The team also continue to receives a small, but valued, selection of postal responses to its problems from senders across the world. All these responses are carefully considered by the team and a selection are published on the website, highlighting a range of approaches toward solving a problem as well as the reasoning behind them. This enables students to share their ideas with a much wider audience than their immediate friends and family, developing both their reasoning skills as well as mathematical communication skills.

NRICH — The early days

Back in the late 1990s, the opportunities afforded by the nascent internet attracted the interest of a team of mathematics teacher educators. At Cambridge's Homerton College (now part of the Faculty of Education, University of Cambridge), teacher educator Toni Beardon envisaged a national online center for curriculum enrichment which would support teachers to promote the teaching and enjoyment of mathematics. From its very beginning, NRICH adopted a pioneering approach; its staff members were recruited from both the Faculties of Mathematics and Education. "The greatest challenge", Beardon explained, "was to create suitable material to meet the monthly deadlines and to reach out across the world to attract young people who would engage in solving NRICH problems and submit solutions" [1]. However, by working with both faculties, Beardon brought together a team of highly experienced classroom teachers, educationalists, and mathematics consultants to design NRICH's earliest mathematics resources. This team worked alongside IT specialists who were tasked to build and further develop NRICH's online platform. They also identified the principles that inform the distinctive design of NRICH activities.

NRICH — Design principles

NRICH activities share certain characteristics that make them instantly recognisable. In this section, we will explore some of those key design principles, beginning with the importance of ensuring NRICH activities have a low threshold and high ceiling.

Low-threshold, high-ceiling activities

NRICH resources are intended to be accessible for students. This makes them ideal for whole-class learning. Their design features tend to incorporate a level of challenge, often unseen at first glance, which enable students to take their learning deeper where appropriate. These features result in NRICH's distinctive resources, which are often described as "low-threshold, high-ceiling" (LTHC) activities. Alison Kiddle designed many NRICH resources and shared her interpretation of NRICH's LTHC approach:

> "The phrase LTHC (also sometimes referred to as 'low-threshold-no-ceiling', or 'low-floor high-ceiling') is believed to originate in Papert's description of the central design principle of the Logo programming language. The 'low threshold' means that new users, including those who have never programmed before, should find it easy to get started, and 'no ceiling' (or 'high-ceiling') means the language shouldn't be limiting for advanced users. We use the term in a similar way when talking about our resources. It can be summed up as follows: a Low Threshold High Ceiling task means *everyone can get started*, and *everyone can get stuck*." [2]

If an NRICH activity is promoted as an LTHC activity for a group of students, then almost everyone in the group would be reasonably expected to at least be able to make a start on it. The threshold needs to be mathematically accessible for all the students in the group, that is, everyone needs to have the prior mathematical knowledge required to start working on the problem. This will of course vary depending on the age and prior attainment of the learners in the group — a task that has a low threshold for 17-year-olds might be far out of the reach of most 7-year-olds!

One problem — Multiple approaches

Solving a problem often involves decision-making because there may be more than one way to approach it. The Chinese mathematics education writer Liping Ma explains why this approach is so important:

> "The reason that one problem can be solved in multiple ways is that mathematics does not consist of isolated rules but connected ideas. Being able to and tending to solve a problem in more than one way, therefore,

reveals the ability and the predilection to make connections between and among mathematical areas and topics." [3]

Clearly, problem-solving is not a "one size fits all" activity. Learning to solve unfamiliar problems means that students must develop a range of problem-solving skills. However, simply knowing a range of skills is insufficient because students must also develop a willingness to work flexibly when appropriate. "Learners should know multiple strategies and demonstrate the critical judgement to choose between them for particular problems" [4]. NRICH activities actively encourage students to work flexibly in several ways. For example, consider the NRICH problem Steel Cables (Figure 1). Due to its LTHC design, its low threshold should enable most students to quickly engage with the problem and begin working out the number of strands in a size 5 cable.

To help students develop a more flexible approach towards problem-solving, NRICH users are often encouraged to explore different approaches through the "Hints" tabs on the problem pages. For example, one of the hints from Steel Cables features a diagram illustrating a possible starting point for students (Figure 2). The use of "Hints" tabs in a problem was an idea based on feedback from students themselves. Through the use of focus groups, NRICH team members found that students often prefer

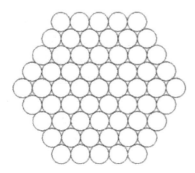

"Cables can be made stronger by compacting them together in a hexagonal formation. Here is a 'size 5' cable. Can you work out, **without counting every strand**, how many strands it contains?"

Figure 1: The NRICH activity, Steel Cables nrich.maths.org/steelcables. Reproduced with permission of NRICH, University of Cambridge, all rights reserved.

Once you've had a go at the problem, click below to see the diagrams some students produced when they worked on it.
Do these diagrams give you any ideas for how you could work out the number of strands needed?

Group 1

Hide

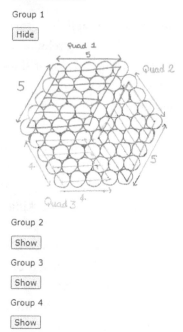

Group 2

Show

Group 3

Show

Group 4

Show

Figure 2: The use of "Hints" to support students exploring the NRICH activity, Steel Cables nrich. maths.org/steelcables. Clicking the "Hide" tab removes the diagram from their view. Reproduced with permission of NRICH, University of Cambridge, all rights reserved.

to choose when to seek support. Students have explained that they feel reassured knowing they can access it when they need to, but can also leave it alone otherwise; they do not like "spoilers". Hence, by adding "Hints" tabs to a problem page, students know that they can easily access a helpful idea to keep them progressing on a problem, but can also ignore it too if they feel like they are making good progress or simply need a little longer to think through their ideas before getting started on a solution. This is a useful approach for helping them to build their mathematical resilience.

Once students have reached a solution, they can compare their approach with a selection of those submitted by other students. At any time, several NRICH problems are presented on the website as "Live" and the team actively encourages students and their teachers to submit their

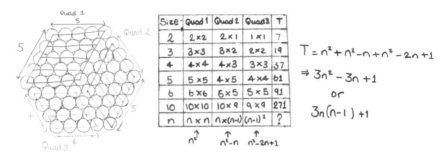

Size	Quad 1	Quad 2	Quad 3	T
2	2×2	2×1	1×1	7
3	3×3	3×2	2×2	19
4	4×4	4×3	3×3	37
5	5×5	4×5	4×4	61
6	6×6	6×5	5×5	91
10	10×10	10×9	9×9	271
n	n×n	n×(n-1)	(n-1)²	?

$$T = n^2 + n^2 - n + n^2 - 2n + 1$$
$$\Rightarrow 3n^2 - 3n + 1$$
or
$$3n(n-1) + 1$$

ideas, adding to the collection of student solutions. The final choice of solutions placed on display is carefully curated from the wide selection sent to the team. Published solutions tend to feature not only the answer to a problem, but also the thinking behind the decisions that have been made along the way, because the journey to reaching that solution is also important. Submitting solutions is very straightforward, most students share their work online through the "Submit a solution" tab on NRICH. The team regularly receives solutions from students from across the world, and the team is always looking out for different approaches toward solving a problem. For Steel Cables, the "Solutions" tab features four very different approaches toward solving the problem. Figure 3 shows the table submitted by the student whose approach was presented as a hint in Figure 2.

NRICH team members often enjoy sharing activities such as Steel Cables with small groups of students during a face-to-face teaching session. Typically, a session might open by allowing the students time to choose their own ways to solve the problem. Then, the session would progress by encouraging them to explore one or more of the other approaches, using partial solutions from other students as prompts. This approach models the importance of valuing flexibility and, as we will explore later in this chapter, is based on academic research. Students need to develop a range of problem-solving strategies and a willingness to choose between them based on the merits of each individual problem. Knowing more than one way to solve problems is also very useful when students inevitably get stuck because, rather than give up, they can simply explore another strategy.

Nurturing mathematical resilience and perseverance

Developing a willingness to "try, try, and try again" is undeniably important for students, they should not give up too easily. However, sometimes our chosen approach is just not suitable for an individual problem. As educators, we need to help students to develop a willingness to explore alternative approaches where appropriate. After all, a problem is only a problem when you do not know the answer right away. Education writer James Nottingham talks about the need for students to face a challenge in order to learn how to cope with adversity and overcome those challenges [5]. When a student gets stuck on an NRICH online problem, they are supported to overcome such a challenge by using the "Hints" tabs for a prompt or exploring the published solutions for more detailed support.

Developing a growth mindset

The idea that students can keep learning new ideas and develop further as mathematicians is a key concept for the NRICH team. Carol Dweck's work on encouraging a growth mindset is a major influence on the work of NRICH [6]. However, not all students arrive in the classroom with a belief that they can improve as a mathematician. For example, they may hold negative attitudes that have been passed down from previous generations. Developing mathematical resilience and addressing mathematical anxiety are becoming increasingly important aspects of NRICH's resource design as more and more researchers are investigating this aspect of mathematical learning. Designing activities where students may experience getting stuck, learn to overcome their difficulties, and reflect on their approaches are all important design issues. The NRICH team actively engages with researchers in this field and contributes toward teacher conferences. Team members also publish articles on NRICH which enable teachers to address these issues with their classes.

NRICH — A research-informed initiative

All NRICH activities are intended to nurture young mathematicians. Their design is informed by educational research, which highlights the need for students develop and embed a wide range of skills. There have been many attempts to identify key skills or mathematical habits in the research literature. Perhaps one of the most influential for the NRICH team is a paper highlighting the need for educators to develop mathematical "Habits of Mind" among their students, such as encouraging them to investigate, spot patterns, and experiment with their ideas [7]. This research inspired the NRICH team to identify and collate existing NRICH activities that encourage students to develop their habits to work collaboratively, think mathematically, ask questions and overcome challenges. These compilations of NRICH activities enable teachers to focus on developing a specific habit through regular practice. Each compilation is organized by curriculum topic, allowing teachers to focus on a specific habit and curriculum area over a sequence of lessons. Alternatively, they may choose to focus on a range of skills during the school year.

As well as influencing the organization of NRICH activities, research also informs their design. NRICH activities go well beyond satisfying the requirements of the school curriculum by addressing the five key mathematical proficiencies proposed by Jeremy Kilpatrick and his team — procedural knowledge, conceptual understanding, adaptive reasoning, strategic competence, and a positive disposition toward learning mathematics [8]. These five proficiencies highlight the importance of encouraging students to justify their ideas and learn how to communicate them to others. They also encourage them to develop a willingness to work flexibly on problems, valuing the use of different approaches where appropriate and encouraging mathematical curiosity as well as problem-solving. Crucially, the proficiencies also encourage students to enjoy their mathematics and develop a mindset required to overcome obstacles when they get stuck. These five proficiencies resonate heavily with the design principles and vision behind NRICH.

To help schools embed these ideas in the classroom, the NRICH team developed a visual aid that encourages the development of the five proficiencies (Figure 4). Using this aid, students working on NRICH activities are encouraged to reflect on their use of the various proficiencies and record their progress visually. This approach supports students to identify their strengths as well as their learning needs, helping them to identify their next learning goals.

Understanding - Maths is a network of linked ideas. I can connect new mathematical thinking to what I already know and understand.

Tools - I have a toolkit that I can choose tools from to help me solve problems. Practising using these tools helps me become a better mathematician.

Problem solving - Problem solving is an important part of Maths. I can use my understanding, skills and reasoning to help me work towards solutions.

Reasoning - Maths is logical. I can convince myself that my thinking is correct and I can explain my reasoning to others.

Attitude - Maths makes sense and is worth spending time on. I can enjoy Maths and become better at it by persevering.

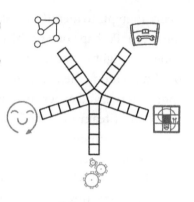

Figure 4: A student-friendly approach toward developing the five proficiencies nrich.maths. org/14652. Based on the model proposed by Kilpatrick *et al.* [6]. NRICH diagram reproduced with permission of NRICH, University of Cambridge, all rights reserved.

Why do this problem?

Many students are accustomed to using number patterns in order to generalise. This problem offers an alternative approach, challenging students to consider multiple ways of looking at the structure of the problem and making sense of other people's approaches, an important part of working mathematically.

The powerful insights from these multiple approaches can help us to derive general formulae, and can lead to students' appreciation of the equivalence of different algebraic expressions.

Figure 5: Extract taken from the Teacher Notes for the NRICH activity, Steel Cables nrich.maths. org/steelcables. Reproduced with permission of NRICH, University of Cambridge, all rights reserved.

Research-informed teacher support

Although NRICH's activities are designed for students, the most effective way to maximize their impact is often by working with teachers. Hence NRICH activities are supported by detailed Teacher Notes. These follow a consistent template for all NRICH activities. The Teacher Notes always begin by outlining the rationale behind the activity, which is helpful for both specialist and non-specialist teachers of mathematics. For example, the Teacher Notes for Steel Cables highlight its potential for considering multiple approaches toward a problem (Figure 5).

All NRICH Teacher Notes highlight key teaching points as well as offering extension ideas and opportunities for support to those who need it. All NRICH Teacher Notes also feature a list of key questions intended to help increase the mathematical potential of the problem in the classroom. These are all consistent design features, appearing on all NRICH Teacher Notes. This consistency saves teachers valuable time since they can focus on delivering the NRICH activities with their classes without spending too much time searching for possible tasks or ways to use them in class. As we saw earlier, the activities are usually accompanied by a range of solutions that teachers can use in a variety of ways, from supporting students to get started on a problem, to challenging others to compare different approaches toward solving the same problem. They are also a very useful aid for encouraging students to develop their own mathematical communication skills and perhaps submit their own solutions to NRICH.

NRICH Teacher Notes reflect the team's research-informed methodology. As mentioned earlier, the approach toward their design is research-informed. NRICH adopts the exploratory learning approach championed by Professor Kenneth Ruthven of the Faculty of Education, University of Cambridge [9]. His three-part approach to the teaching and learning of mathematics, "exploration, codification, consolidation", is a corner stone of the team's approach to resource development and the dissemination of its ideas. This model of teaching mathematics involves an initial exploration phase, followed by teacher codification to draw together the ideas discovered by students in the exploratory phase, before moving on to consolidation work. Although this approach differs markedly from the "model, practice" approach seen in many schools, Ruthven's approach was adopted for the project because it enables students to experience mathematics as a journey of discovery.

NRICH — Building a worldwide audience

At the time of writing, NRICH has a worldwide audience of millions of teachers, students, and their families. Such a huge, sustained audience is the outcome of many years of hard work. Nowadays, social media can enable a new teaching resource or idea to attract the attention of millions within an incredibly short time span, although such ideas may also quickly fall out of favor. However, NRICH has consistently attracted large audiences over two decades; the team took a very sustainable approach toward building its audience, relying heavily on recommendations from teachers, lecturers, and professional development leaders. This approach worked well because teachers are well-known for sharing high quality resources with one another, enabling NRICH resources to become quickly adopted by many teachers. The NRICH team started attending mathematics subject association meetings, volunteering to lead sessions using NRICH activities to model their use, and further increase awareness about NRICH. Team members also worked closely with the mathematics advisors for local areas, knowing that they would be very likely to share NRICH activities with their local schools.

Despite the internet being very much in its development phase when NRICH started out in the late 1990s, its early success quickly resulted in schools sending in solutions from their classes. "We were never short of good solutions from students," explained resource designer Bernard Bagnall, "Some sent independently, some after work in school." The team published some of those student solutions on NRICH, attracting even more visitors to its website, "We were delighted that the number of hits steadily increased," reflected Bernard. Two decades later, the NRICH team continues to welcome solutions from students enjoying the team's latest activities. More recently, social media has offered another platform for the team to share ideas and feedback with schools and their students. During the recent COVID-19 crisis, the value of NRICH's freely available activities and support materials was demonstrated by the massive number of schools and families turning to NRICH for support; the team welcomed over 1.25 million pageviews each week following the school closures (*Source*: Google Analytics).

Embedding Problem-Solving in the Classroom

Although problem-solving is generally regarded as an extremely important skill to develop, embedding any resources in schools depends highly on the motivation of individual teachers as well as their school's mathematics policy. Some schools choose to set aside 1 day a week for problem-solving, others identify a problem-solving week in their annual calendars. However, problem-solving requires many different skills, which demands regular practice and reflection. To enable more students to enjoy problem-solving activities more often, NRICH activities are mapped to the mathematics curriculum. This approach enables teachers to identify at least one suitable NRICH activity for almost every mathematics lesson that they teach. For example, Figure 6 illustrates the range of NRICH problems suitable for supporting Year 1 students (5- and 6-year-olds) to develop specific, curriculum-linked counting skills.

For the objective "Count, read and write numbers to 100 in numerals; count in multiples of twos, fives and tens," teachers visiting the NRICH curriculum map will find five highly suitable activities. By matching NRICH activities with the curriculum, teachers are encouraged to regard every lesson as an opportunity to develop problem-solving skills, knowing that their chosen activity will enable them to cover the required curriculum content too.

Reducing teacher workload is a key ambition of the NRICH team. Activities on the curriculum map are labeled to provide further details about their level of challenge and activity type, an approach intended to reduce teacher workload searching for suitable activities for their classes. Each activity has a star rating which indicates its level of challenge. A one-star activity (such as Biscuit Decorations in Figure 6) is generally regarded as suitable for introducing to a whole class whereas a two-star activity indicates that it is suitable for the majority of students, three-star

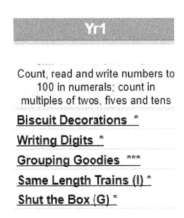

Figure 6: Screenshot taken from NRICH's curriculum map for Year 1 students NRICH curriculum map for 5–11 years old. Reproduced with permission of NRICH, University of Cambridge, all rights reserved.

activities suggest that it offers a greater level of challenge (such as Groupling Goodies in Figure 6).

The curriculum map also supports teachers by identifying the type of activity that each resource can offer their class. These may be games (such as Shut the Box in Figure 6) or an investigation (such as Same Length Trains in Figure 6).

Maximizing the Potential of NRICH Resources

Although NRICH provides detailed Teacher Notes and free curriculum maps for its activities, modeling their use is also very important. NRICH does this in different ways. The team organizes face-to-face training sessions for teachers in the UK and further afield. These sessions enable NRICH team members to model the three-part lesson approach and encourage teachers to reflect on the approach for their own classes. More recently, the team has explored projects intended to increase parental engagement with mathematics. This led to the team revisiting its use of Teacher Notes to support its activities. For parents, the team realized that recording short 2- or 3-minute videos modeling the use of an activity would provide families with the necessary support in a very efficient way. More recently, the team has been exploring the use of webinars to model the effective use of NRICH activities.

Another approach for modeling the use of resources involves students. The NRICH team has developed a roadshow that visits schools to share a range of hands-on NRICH activities throughout a day. These sessions are very popular and soon get booked up, enabling NRICH colleagues to model the use of their activities with classes as their teachers observe the sessions. Further support is provided by uploading the resources to NRICH, alongside Teacher Notes and solutions. The team has also found it beneficial to organize a staff meeting either before or after a roadshow visit to draw out key aspects of the NRICH approach.

Next Steps and Future Opportunities

Students being able to access high quality, research-informed mathematical activities has always been important. This was further highlighted during the COVID-closures; far too many students received a very restricted education, if at all, and many were reduced to repetitive practice of procedures and rapid recall activities. NRICH offers opportunities for students to engage with activities that nurture a much wider range of skills and habits, and the website encourages students to share their solutions and compare them with those of other students.

We know that our students will be entering a working environment where problem-solving and teamwork skills will become increasingly important as more and more roles become automated. Going forwards, NRICH will focus its efforts on further developing the aspects of its work that enable students to develop those key skills and habits. That way, more students will be equipped to thrive in their future lives, whatever happens in our rapidly changing world.

References

[1] Beardon, T. (2016). *NRICH @ 20*. Primary Mathematics: The Mathematical Association.

[2] Kiddle, A. (2019). *Low Threshold, High Ceiling — an Introduction*. Accessed from nrich.maths.org/10345.

[3] Ma, L. (1999). *Knowing and Teaching Elementary Mathematics: Teachers' Understanding of Fundamental Mathematics in China and the United States*. Mahwah, NJ: Lawrence Erlbaum Associates, Inc.

[4] Star, J. R., Rittle-Johnson, B., and Durkin, K. (2016). "Comparison and explanation of multiple strategies: One example of a small step forward for improving mathematics education". *Policy Insights from the Behavioral and Brain Sciences*, 3(2), 151–159.

[5] Nottingham, J. (2008). "Challenging our thinking; Challenging our schools". In *The 13th International Conference on Thinking Norrköping, Sweden June 17–21, 2007* (No. 021). Linköping University Electronic Press.

[6] Dweck, C. S. (2006). *Mindset: The New Psychology of Success*. New York: Random House.

[7] Cuoco, A., Goldenberg, E. P., and Mark, J. (1996). "Habits of mind: An organizing principle for mathematics curricula". *The Journal of Mathematical Behavior*, 15(4), 375–402.

[8] Kilpatrick, J., Swafford, J., and Findell, B. (2001). *Adding It Up: Helping Children Learn Mathematics*. Washington, DC: National Academies Press.

[9] Ruthven, K. (1989). "An exploratory approach to advanced mathematics". *Educational Studies in Mathematics*, 20(4), 449–467.

© 2023 World Scientific Publishing Co. Pte. Ltd.
https://doi.org/10.1142/9789811253072_0003

Chapter 2

Mathematical-Artistic Activities for Social Inclusion and Well-Being: The Experience Workshop STEAM Network

Kristóf Fenyvesi*,†,¶, Christopher Brownell*,‡,**, Osmo Pekonen*,§,††,
Zsolt Lavicza*,‖,‡‡, and Nóra Somlyódy*,§§

Experience Workshop, Finland
†Finnish Institute for Educational Research, University of Jyväskylä, Finland
‡Fresno Pacific University, USA
§University of Jyväskylä, Finland
‖Johannes Kepler University, Austria
¶fenyvesi.kristof@gmail.com
***chris.brownell@fresno.edu*
††osmo.pekonen@jyu.fi
‡‡lavicza@gmail.com
§§nora.somlyody@experienceworkshop.org

The Experience Workshop — International STEAM Network started in 2008 in Hungary and 2010 in Finland as a collaborative effort of mathematicians, artists, teachers, parents, and children. Experience Workshop has not formalized its' membership. Members are loosely connected through various mailing lists and social media groups, and they are involved in a variety of events, conferences, and workshops in the role of contributors or participants. Experience Workshop organizes various math-art-education events and interactive exhibitions, workshops, seminars, and trainings all over Europe, along with educational and research collaborations in Africa, America, Asia, and Australia. Over the years, more than 40,000 people participated in our projects. Connections between mathematics and arts education provide Experience Workshop's pedagogical basis. As mathematics and arts connections always involve a wide cultural background, we are increasingly focused on the need for Science, Technology, Engineering, Arts, and Mathematics (STEAM) integration in learning. The development of collaborative and inter-, multi-, and transdisciplinary problem-solving capabilities, which enable students to discover unexpected connections between different aspects of various complex phenomena, is not only a useful tool but is recognized as one of the most important goals of today's learning societies. STEAM in today's European schools represents a dynamically developing, but mostly unexplored field. In this chapter, we introduce our methods and various tools from Experience Workshop's collection. By introducing one of our projects in detail, we would like to highlight especially those components focused on "mathematical arts and creativities", uniting these diverse educational cultures with a more global perspective.

Introduction: Experience Workshop's Goals

Experience Workshop[a] was launched as the collaborative effort of mathematicians, artists, teachers, parents, and children in 2008 in Hungary. The community has grown out from the Ars GEometrica mathematics and art conferences held between 2007 and 2010 in Hungary. It was initiated by education researcher Kristóf Fenyvesi, mathematics and physics teacher Ildikó Szabó, and the mathematical artist, Dániel Erdély. The goal of the community was to introduce various math-art innovations with educational potentials to basic and early childhood education. The Experience Workshop was an independent organization from the beginning, with the intention of financing its activities from regional and European Union projects, and later, from product sales and event and training fees. Over the years, Experience Workshop has organized hundreds of math-art-education events, exhibitions, workshops, seminars, and training programs worldwide. Thanks to the members' openness toward each others' innovative activities in education, Experience Workshop grew relatively quickly into a vast network of professionals and enthusiasts, offering a diverse set of high quality and well-tested educational implementations of math-art connections.

Our members share common interests of developing an integrated approach to STEAM content across various levels of education. STEAM became a popular educational concept in the 2010s [1] as an extension of STEM integration with the arts. STEAM inherited the difficulties of defining STEM, with the added complication of a lack of clarity related to the meaning and role of the "A" [2], arts. STEAM is mostly associated with 21st century skills and competence-based education and inter-, cross-, multi-, and transdisciplinary approaches, implemented in project-based frameworks to go beyond the traditions of subject-based learning in education. John Maeda in the Rhode Island School of Design successfully coordinated the STEM to STEAM movement to include design thinking, creativity, and innovation [3] in educational policies around the globe.

Experience Workshop's current international network involves teachers, artists, scholars, artisans, designers, inventors, and producers of educational

---

[a] Experience Workshop's website: www.experienceworkshop.org [Accessed on April 28, 2021].

Figure 1: Results of Experience Workshop's hands-on modeling activity of Fullerene "balls".

toolkits. Since 2010, the organization has been active in Finland as well and is coordinated jointly from Hungary and Finland. Participants of Experience Workshop's programs are able to learn mathematics through the arts and make art with mathematics in a playful manner (Figure 1). The community aims to initiate and collaborate in local and international projects. Several international projects are funded by the European Commission. These involve the development and distribution of new mathematical-artistic toolkits for basic education, such as the Polyuniverse [4] or Logifaces (see section "Building bridges with STEAM education toolkits"). There are other projects, which are providing new curricula and didactics. Experience Workshop, in collaboration with the influential mathematics popularizer community IMAGINARY and other international partners, recently released Mathina (https://mathina. eu/), an interactive storybook connecting mathematical problem-solving with imagination and storytelling. In the Hallå STEAM project, engaging mathematical-artistic activities are combined with Swedish lessons in Finnish schools. This program implements live role-playing, where researchers of Finnish-Swedish historical connections in science, Osmo Pekonen and Johan Stén, often appear in 18th century costumes and engage the students in various problem-solving activities in the spirit of the Age of Enlightenment [5]. These projects explore the possibilities of innovative didactics and educational programs involving children, parents, and teachers in a vivid exchange between the mathematical and artistic perspectives of looking at the world.[b]

In the context of formal education, our projects span K-12 education. Experience Workshop develops math-art resources for teachers, students, and parents like books, collections, methodological materials, and scientific articles, a lot of which is open-access.[c] Experience Workshop's research encompasses various pedagogical methodologies in connection with STEAM education, such as project-based [6,7], inquiry-based [8], collaborative [9], playful [10], and experience-oriented [11] mathematics learning; connections between hands-on activities, digital modeling, and fabrication; science and art combinations; phenomenon-based and multidisciplinary education.

Although Experience Workshop's activities are continuously broadening and transforming, their primary goals have not changed since the beginning:

- Integrating art-, creativity-, and play-oriented activities into mathematics learning and playfully integrating mathematics in learning other subjects, especially the arts.

- Implementing math-art and STEAM approaches into real-world problem-solving as part of the learning process.

- Organizing math-art and STEAM events to introduce best practices in the experience-oriented teaching of mathematics (Figure 2).

- Making students, future, and in-service teachers familiar with the most recent results (including didactics, toolkits, resources, etc.) of experience-oriented mathematics education; researching, collecting, and publishing outstanding achievements in the field and making these accessible for the academic, artistic, and educational communities.

- Expanding educators' everyday collection of innovative educational approaches through toolkits to increase learners' mathematical, logical, combinatorial and spatial skills, computational thinking, developing perception, and aesthetic sensibility, thus motivating individual and

---

[b] See Experience Workshop's main current international projects here: https://www.experienceworkshop.org/steam-education/?lang=en [Accessed on April 28, 2021].

[c] See some of Experience Workshop's resources here: https://www.experienceworkshop.org/resources/?lang=en [Accessed on April 28, 2021].

Figure 2: Experience Workshop's STEAM playground at Helsinki Design Festival in 2019.

collaborative problem-solving, interdisciplinary and inter-artistic approaches on all levels of education.

Experience Workshop's Methodological Background

Both the advocates of problem-solving in mathematics pedagogy and the pioneers of art and science connections in aesthetics have emphasized that systematic experimentation with establishing connections between different components of a problem lead to unexpected solutions. From its inception, Experience Workshop's members have been urged to rediscover those mathematicians, artists, and theorists whose work can support the renewal of education from a mathematical-artistic perspective. In the Hungarian context, Experience Workshop has taken its inspiration from the didactic works of György (George) Pólya, Rózsa Péter, Zoltán Pál Dienes, and Tamás Varga in the field of problem-solving and the development of mathematical and computational thinking, along with László Moholy-Nagy, György (George) Kepes, Victor Vasarely in art theory, philosophy, and art education practices, which emphasize the relationship between art and science. These form the core of Experience Workshop's holistic pedagogical approach.

In addition to the "Hungarian traditions" of problem-solving in mathematics and interdisciplinary approaches in the arts, when it comes to the holistic, competency-based, inter-, and multidisciplinary pedagogical approaches, Experience Workshop draws intentionally on the Finnish National Core Curriculum [12]. An essential feature of Experience Workshop's approach is project-based learning and inclusive pedagogy, requiring small groups' collaboration. The learning content is primarily based on the participants' creative activities, as they are, individually as well as on a group level, involved in the planning and evaluation of the learning process.

Participants of Experience Workshop's programs actively influence the playful learning process through their interests and preferences. In line with the main recommendations of the Finnish National Core Curriculum [12], they are encouraged to reflect on what they are learning, to express their emotions, interact with other participants, and transform their physical and mental environment — which also plays an essential role in the learning process.

The crucial outcomes of learning in Experience Workshop's programs are to develop creativity, critical thinking, and problem-solving skills as

Figure 3: Collaborative problem-solving with Polyuniverse toolkit from Experience Workshop's collection.

Figure 4:   Symmetry Workshop with Carawonga from Experience Workshop's collection.

well as to exercise negotiation with others in collaborative problem-solving processes (Figures 3–4). With these, they can then consider and find value in external perspectives and understand the views of other participants. The discovery of mathematics' multivarious roles in our world, with a special emphasis on arts, and the activity-based exploration of artistic connections in mathematics, provides an inspiring framework for the realization of these complex aims.

It is not possible to introduce Experience Workshop's various activities in a single article. Therefore, we decided to introduce a project that highlights the effectiveness of mathematical-artistic activities in a rather special context: learning for social inclusion.

Mathematics and Art Connections for Social Inclusion: The Sillat Project in Finland

Between August 2018 and June 2020, Experience Workshop employed the STEAM framework to support the social integration of young immigrants in Finland. Backed by the Finnish National Board of Education's funding, the Finnish NGO organization called Settlement of Jyvälä was the main coordinator of the project, and Experience Workshop provided the pedagogical program based on math-art connections in STEAM context. Settlement of Jyvälä ran an adult education center, organized after-school activities and civic programs for children, provided children, youth, and family services based on a comprehensive experience of working with immigrant youth and families as well.

The project was called "Sillat" in Finnish, which means "Bridges", and it had two priorities:

- Supporting the integration, inclusion, and well-being of young immigrants through Experience Workshop's math-art methodology and STEAM approach.

- Promoting good cultural relations by increasing opportunities for encounters and dialogue between those already living in Finland, and those who have moved recently to Finland from elsewhere. This latter refers to a very heterogeneous group of international students, asylum seekers, expats, etc.

The basic structure of the project was (1) group activities with young adult immigrants who had recently moved to Finland, (2) public events open to local families and anyone else, and (3) workshops for teachers, youth workers, and other specialists, mainly from the social sector. As part of the group activities, Experience Workshop's facilitators involved the Sillat project participants in STEAM projects. In the process of learning-by-doing, they became familiar with the goals of Experience Workshop's learning programs, including specific topics in mathematics and the arts, and the materials and toolkits that are used. During the project Experience Workshop's facilitators also became familiar with Sillat participants' interests and backgrounds, which were reflected in the group activities. Our goal was to equip participants with mentoring experience and a related set of soft skills, and support their self-confidence to become STEAM education facilitators themselves at public events and at the professional workshops offered by the Sillat project. The public events were mainly connected to the main social and cultural events of Jyväskylä city, and enabled encounters between people from different backgrounds in a context where the shared STEAM content and Experience Workshop's math-art toolkits and playful problem-solving activities became a bridge for communication. In addition to these activities, Experience Workshop provided STEAM training programs for teachers, tutors, and peer tutors working with immigrants.

STEAM for inclusive learning communities and well-being

Experience Workshop implemented the integrated approach to STEAM in the Sillat project. STEAM was useful in creating links to the Finnish National Core Curriculum, which recommended multidisciplinary projects for developing transversal skills and student-centered, multidisciplinary, and phenomenon-based learning methods. The STEAM approach also supported applying these methods in a youth and adult learning context, where inclusion played a critical role. In Experience Workshop's STEAM projects, it was essential to introduce different perspectives to motivate all learners. STEAM proved to be an effective pedagogical tool to achieve the socially and culturally inclusive goals of the project.

In the Sillat project, the STEAM approach also played an essential role in enhancing community work, a critical component in our program. As part of the collaborative problem-solving activities, all participants recognized each other and themselves mutually as creative and skilled individuals with diverse competencies from which others could learn. To share their knowledge and skills, participants first had to discover and identify their own strengths to build self-confidence. They gained motivation through engaging activities, developed their social competencies, and recognized the values of creative exchange. During the workshops and public events, participants developed their communication and mentoring skills, picked up the Finnish language, and got familiar with local culture, including people's attitudes, children's interests and many more details. Meanwhile, they could also introduce their cultural and personal backgrounds to each other and to locals participating in the Sillat project's public events and professional workshops.

Group activities

The fundamental principles in designing the group activities of the Sillat project were:

- Developing collaborative creativities and community skills through joint problem-solving sessions using math-art content.
- Working together, so that everyone had the opportunity to participate according to their personal skills.
- Providing appropriate environments, personal, and community spaces, time, support and occasions to succeed.
- Ensuring low-threshold participation for verbal activities, in consideration of group activities participants who were Finnish language learners.

The 2-year-long Sillat project included several programs. On average, there were two to three different activities monthly. We have chosen to introduce only those in this chapter that focused on mathematics and arts connections, and most characteristically represented the integrative potential of STEAM. Most of these programs focused on the transdisciplinary phenomenon of symmetry, and contributed to social inclusion and development of thinking skills through math-art connections.

Building bridges through music and mathematics

Music can play an essential role in building bridges between people. Music in the Sillat project brought participants the experience of joy, togetherness, and mutual support. Music also provided support for hands-on math-art activities (Figure 5). Moreover, developing thinking skills by playing with musical patterns, where musical symmetries often combined with symmetric movement patterns (see section "Public Events" and Figure 14), was a frequent warm-up or ice-breaking community activity played by the project participants. Exploring musical symmetries in traditional musical patterns by clapping, and in traditional dances by counting moves and directions of steps was a frequent activity in the Sillat project.

Figure 5: Singing together inside Experience Workshop's Geodesic Dome as the finale of a STEAM activity.

Building bridges with STEAM education toolkits

We have used a wide variety of STEAM education toolkits from Experience Workshop's portfolio to stimulate creative thinking, enhance problem-solving, strengthen collaboration skills, and grow team spirit. This section includes only a few examples from the wide collection of various programs, mainly those that had direct connections to the exploration of various symmetrical phenomena and are based on some of these toolkits, which might be less known even in the math-art discourse.

Pentiles by Haresh Lalvani. Invention and fun — All of them connect!
Creating new patterns with the New York-based Haresh Lalvani's Pentiles turned out to be an interesting and meditative experience for Sillat project's participants and the visitors of the public events. Pentiles (as the Greek word for the number 5, "penta-", hidden in the name of the toolkit also signifies is based on the number 5, and all tiles have angles related to the pentagon (5-gon) and decagon (10-gon)). All tiles fit with each other in a "repeating" (periodic) and "non-repeating" (aperiodic) manner, where

some could be tiled by themselves, while others required pairs or more shapes to tile the surface.

The designer of Pentiles is Professor Haresh Lalvani, the New York-based artist, designer, scholar, and the Director of the Center for Experimental Structures, at Pratt Institute. Lalvani invented Pentiles between 1981 and 1991, and it was the subject of four patents and appeared in parts in several publications (Figure 6). Applications to products — carpet tiles, puzzles, and textiles — followed in the 1990s. However, efforts to introduce them as children's play blocks were yet to come... Among several international high visibility events, Experience Workshop successfully implemented Pentiles in the Sillat project, and the participants came up with unique compositions and had the opportunity to study various symmetries in a learning-by-doing manner (Figure 7).

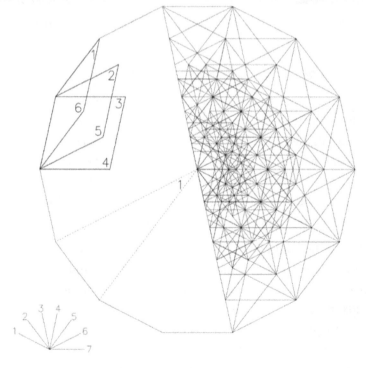

Figure 6:   Haresh Lalvani's Pentile system from Experience Workshop's collection.

Agathias GianTiles. Playing with a large wooden math-art tessellation toolkit, created by Gábor Gondos. The Agathias tessellation is made of large wooden tiles. The diameter of the assembled tessellation is 180 cm. The visual inspiration for the unique pattern came from an ancient codex dating from 1483, called *Agathias* from the *Bibliotheca Corviniana*, Matthias Corvinus' famous Renaissance collection. The Hungarian math-artist, Gábor Gondos, was inspired by how M.C. Escher was using the triangular tile design in an inventive, figural way to obtain the organic shapes of every single tile. After several experiments, Gondos developed the tessellation into an educational toolkit and cut the tiles with a laser-cutter from plywood.

Assembling the Agathias puzzle was a collaborative problem-solving activity, which required logical planning that implemented combinatorics, algorithmic thinking, and the ability to recognize symmetrical configurations in the emerging composition. To solve the

Haresh Lalvani's Pentiles @ Jyväskylä, Finland. Experience Workshop STEAM Network. www.experienceworkshop.org

Figure 7: Haresh Lalvani's Pentiles from Experience Workshop's collection at a Sillat project group activity.

Figure 8: Assembling Gábor Gondos' Agathias puzzle from Experience Workshop's collection.

problem, participants needed to support each other, look at the pattern from different perspectives, and coordinate their actions on the community level (Figure 8). This provided learning opportunities within and beyond the horizons of mathematical connections in art.

CaraWonga magic carpet to experience symmetries and visual illusions. CaraWonga is a modular carpet made of 80% recycled industrial felt. From the modular pieces of various colors, the participants can create their own carpet designs. CaraWonga's modular system is based on Rhombille tiling, an excellent source for creating complex symmetric patterns, including three-dimensional visual illusions in a simple and intuitive way. CaraWonga is created by the Hungarian textile designer Katalin Gál. A unique feature of the CaraWonga carpet is that it can be assembled and disassembled countless times, based on its novel folding system. There is no one right way to combine the modules, but you can either try to re-create patterns that you have seen or come up with your own unique pattern. When CaraWonga carpets were assembled as part of community activities, both the design and the realization involved calculation and algorithmic thinking skills (Figure 9).

Figure 9:   Assembling the CaraWonga carpet from Experience Workshop's collection.

Assembling the CaraWonga "magic" mosaic carpet had several positive effects. Collaboration was needed during the design and assembly. In the process of assembly, it was natural to work in pairs or smaller sub-groups. Cooperation didn't even require a common language. Participants experimented by showing, explaining their ideas, and communicating through gestures when searching for the right color pieces. The assembly encouraged collaboration and promoted new kinds of interactions within the group. The result achieved together strengthened the community experience and established a new sense of success, especially when the Sillat group's first large masterpiece was exhibited in the lobby of the cultural center, where Sillat meetings took place, and it was admired by all visitors of the building for months (Figure 10).

The logic of assembling the carpet modules proved to not be exceedingly difficult, neither self-evident. Finding the solution always encouraged interaction and at public events, when Sillat participants appeared in mentoring roles, made strangers communicate with each other. Finding a solution challenged logical and combinatorial thinking as well as collaboration skills. Various types of symmetries were possible to discover through the activities. Everyone, regardless of language skills

www.experienceworkshop.org

Figure 10: Sillat's CaraWonga carpet from Experience Workshop's collection exhibited in Settlementi, Finland.

and formal educational background, could experience success. Assembling the CaraWonga carpet turned out to be a great way to make art together and develop collaborative creativity, as well as to develop the concrete vocabulary associated with making the carpet (e.g., colors, numbers, directions, verbs related to assembling, and words of encouragement to bring the teamwork forward). The carpet assembly was a dynamic activity that also changed the use of the learning environment.

Logifaces: Embodied learning with a hands-on math-art toolkit. The Logifaces toolkit successfully inspired playful and embodied learning experiences and offered apparent connections between mathematical problem-solving, design- and art-based challenges along with symmetry observations for the Sillat group. The toolkit proved its potential to develop spatial, visual, and computational thinking, and engaged project participants in a multidisciplinary learning project to explore specific phenomena from both scientific and artistic perspectives, and to improve key competencies.

Logifaces is an award-winning geometric puzzle from Experience Workshop STEAM Network's educational collection. Logifaces implements three-dimensional shapes to provide logical and creative challenges through simple problem-solving activities for various educational contexts and age groups. Currently, an international group of educational experts, including teachers of different subjects, researchers, and designers from Austria, Finland, Hungary, and Serbia, are collaborating on the "Logifaces: analog game for digital minds" Erasmus+ project funded by the European Commission to extend the applications of the game from mathematics to further domains. Experience Workshop represents Finland in this project.

A Logifaces set includes either nine or sixteen modules, consisting mostly of truncated triangular prisms and a few regular triangular prisms. Each truncated triangular prism is distinguishable through the height of the three vertices that can be one, two, or three units. However, this property is not communicated with the players; they can find out these characteristics through systematic exploration of the modules' properties in the set. The goal is to arrange the modules next to each other to build a continuous surface without disruptions. Even though it sounds simple, the game is logically quite demanding. Besides these instructions, the Sillat project's participants were encouraged to invent their own games and to create their own geometrical art with the sets.

As an introduction activity, Sillat participants tried to create animal shapes from Logifaces modules: a cat, a snake, a butterfly, a fish, etc. Pattern

Figure 11: Wooden and concrete Logifaces blocks from Experience Workshop's collection in action at Helsinki Design Festival, 2019.

recognition, tactile, and visual perception were involved in this kind of problem-solving activity. The related Finnish vocabulary of the participants was developed simultaneously as a natural part of this math-art activity. As a next step, three-person student groups modeled Logifaces modules by body movement. Matching the literally embodied Logifaces modules became a real, embodied learning experience in an engaging, playful, and collaborative problem-solving framework. In addition to the community building aspects, this activity helped to open new perspectives and the participants gained a deeper understanding of the geometric shapes (Figure 11).

Public events

As a continuation of group activities, Sillat participants could implement their newly gained math-art and STEAM knowledge and test their growing mentoring skills at several public events. We integrated most of the Sillat public events into city festivals and other significant happenings open to a general audience, including families with children. At these events, the Sillat participants acted as workshop leaders, and they showcased their mentoring skills in a real-life situation. There were more than ten public events in the 2 years of the program, and most of them were part of the largest city festivals. We would like to mention only those where math-art had a central role in the facilitation of the program.

At the City of Light Festival in Jyväskylä and the Helsinki Design Festival, the Sillat project's participants conducted the joint construction of a five-meter tall model of the Warka Tower (Figure 12) as part of a collaborative problem-solving process. The tower consisted of 225 plastic pipes and connectors. The Warka Tower was initially designed to help the Ethiopian people to condense water from air. We used our version of the Warka Tower to "harvest" great experiences in a culturally diverse community [13].

First, the team of Sillat participants began building the tower. The construction process was open to all visitors of the festival, and gradually more and more enthusiastic people — children, young adults, and seniors — joined the crowd to build. The construction required collaboration between people who previously did not know each other. The Sillat participants steered and mediated the process. The Warka Tower rose high based on the intuitive exploration of its' symmetric patterns.

Figure 12: Warka Water Tower from Experience Workshop's collection at the City of Light Festival in Jyväskylä, Finland, as part of the Sillat project in 2019.

As evening fell at the City of Light Festival, the highly symmetric structure transformed into an extraordinary light artwork. The people gathered to admire and dance around the tower. A young Afghan Sillat participant commented that this was the first time he was celebrated for something based on his skills.

At the math-art playgrounds, which we provided at the public events, children were supported by the Sillat mentors and enjoyed construction activities with various educational toolkits.

Embodied Creativities and Embodied Mathematics: Math-Art in Action in the Sillat Project

The Sillat project also contributed to the development of transversal competencies in an informal, collaborative, and creative framework. In addition to motivation and engagement, the several math-art workshops included in this project provided several practical examples of physical

Figure 13: Sillat Workshop for special educators and social workers: "Together we are more! Community Fractal". 4D frame from Experience Workshop collection can be easily rendered into a Sierpinski pyramid. The exercise shows how collaborative problem-solving supports success.

Figure 14: Embodied Creativities for Maths in Motion. Sillat Workshop for special educators and social workers; in this exercise, participants experienced how body movement and dance can be implemented to teach abstract concepts, even mathematical ones.

and embodied experience's connections to mathematical concepts in the field of mathematics (Figures 13–14). According to research, this has several consequences to mathematics learning (e.g., [14–16]). Theories on the relationship between bodily schemas and cognitive domains [17], experiences and linguistic forms [18], gesture and speech [19], spatial/geometrical concepts, and the multiplicity of perceptual experiences based on body movements [20] contribute to the sophisticated interpretation of mathematical conceptions.

Although Sillat was not primarily a mathematics learning project, from the perspective of problem-solving and the development of algorithmic thinking skills, it was worth considering verbal, tactile, and other perceptual interactions, gestures, spatial movements, and object manipulations in different configurations for conceptual understanding [21]. Because of the limited Finnish language skills of the Sillat participants, we needed to invent and collect various activities, which shifted the emphasis from the verbal articulation of knowledge to the participants' bodily engagement in the learning process [22]. The implementation of manipulating objects, modeling, drawing, measuring, physical movement, and other embodied performative actions as part of individual and collaborative problem-solving activities in the Sillat project, supported the interplay between the verbal and non-verbal modes of learning in general and especially in mathematical knowledge development. Since the Sillat project's participants were often lacking a common language, the universality of mathematics built some sturdy bridges they could walk upon.

Concluding Remarks

During the Sillat project, many seeds began to germinate and bear fruit. The participants of the project gained new perspectives and experiences of empowerment; they were able to test their skills in new contexts and experiment with new ways of working in a creative ecological framework. The Sillat project was received very well by the participants and it hopefully provided a vivid example of a certain type of Experience Workshop's activities. At the same time, Sillat also represents a special case of how math-art and STEAM can be implemented for the establishment and the development of a learning community for children and youth with diverse social backgrounds.

Acknowledgments

The authors would like to thank every staff member of the Settlement of Jyvälä and the participants and coordinators of the Sillat project, especially Maia Fandi, Maria Typpö, Ismael Sheikh Musse, Sharifa Sarkhosh, Khaled Aden Yusuf, Anni Niemelä, and Mickaël Fandi.

References

[1] Yakman, G. (2010). *STE@M Education: An Overview of Creating a Model of Integrative Education.* http://steamedu.com/wp-content/uploads/2014/12/2008-PATT-Publication-STEAM.pdf [Accessed on June 1, 2021].

[2] Colucci-Gray, L., Burnard, P., Cooke, C., Davies, R., Gray, D., and Trowsdale, J. (2017). Reviewing the potential and challenges of developing STEAM

education through creative pedagogies for 21st learning: How can school curricula be broadened towards a more responsive, dynamic, and inclusive form of education? BERA Research Commission. https://doi.org/10.13140/RG.2.2.22452.76161.

[3] Maeda, J. (2013). "STEM + Art = STEAM". *STEAM*, 1(1), 1–3. https://doi.org/10.5642/steam.201301.34.

[4] Fenyvesi, K., Osborne, K., Kaukolinna, M., Sinnemäki, M., Kuorikoski, L., and Somlyódy, N. (2020). "Multidisciplinary symmetry education with poly-universe toolkit in schools and informal learning context in Finland". *Symmetry: Culture and Science*, 31(1), 23–34. DOI: 10.26830/symmetry_2020_1_023.

[5] Pekonen, O. and Stén, J. (2020). "Hallå STEAM: Building bridges between mathematics, arts, and humanities". In C. Yackel, E. Torrence, K. Fenyvesi, B. Bosch, C. S. Kaplan (eds.), *Bridges 2020 Conference Proceedings* (pp. 475–478). Tessellations Publishing.

[6] Capraro, R. M., Capraro, M. M., and Morgan, J. R. (eds.) (2013). *STEM Project-based Learning: An Integrated Science, Technology, Engineering, and Mathematics (STEM) Approach*. Springer Science & Business Media.

[7] Kokotsaki, M. V. and Wiggins, A. (2016). "Project-based learning: A review of the literature". *Improving Schools*, 19(3), 267–277.

[8] Viilo, M., Seitamaa-Hakkarainen, P., and Hakkarainen, K. (2018). "Long-term teacher orchestration of technology-mediated collaborative inquiry". *Scandinavian Journal of Educational Research*, 62(3), 407–432.

[9] Laal, M. and Ghodsi, S. M. (2012). "Benefits of collaborative learning". *Procedia-Social and Behavioral Sciences*, 31, 486–490.

[10] Kiili, K., De Freitas, S., Arnab, S., and Lainema, T. (2012). "The design principles for flow experience in educational games". *Procedia Computer Science*, 15, 78–91.

[11] Scogin, S. C., Kruger, C. J., Jekkals, R. E., and Steinfeldt, C. (2017). "Learning by experience in a standardized testing culture: Investigation of a middle school experiential learning program". *Journal of Experiential Education*, 40(1), 39–57.

[12] Finnish National Agency for Education (FNAE) (2016). *National Core Curriculum for Basic Education 2014*. Helsinki.

[13] Fenyvesi, K., Park, H.-G., Choi, T., Song, K., and Ahn, S. (2016). "Modelling environmental problem-solving through STEAM activities: 4D frame's Warka Water Workshop". In E. Torrence, B. Torrence, C. H. Séquin, D. McKenna, K. Fenyvesi, and R. Sarhangi (eds.), *Proceedings of Bridges 2016: Mathematics,*

Music, Art, Architecture, Education, Culture (pp. 601–608). Bridges Finland, Phoenix: Tessellations Publishing. http://archive.bridgesmathart.org/2016/bridges2016-601.pdf.

[14] Nemirovsky, R. and Ferrara, F. (2009). "Mathematical imagination and embodied cognition". *Educational Studies in Mathematics*, 70(2), 159–174.

[15] Núñez, R., Edwards, L., and Matos, J. (1999). "Embodied cognition as grounding for situated and context in mathematics education". *Educational Studies in Mathematics*, 39(1–3), 45–65.

[16] Radford, L. (2009). "Why do gestures matter? Sensuous cognition and the palpability of mathematical meanings". *Educational Studies in Mathematics*, 70(2), 111–126.

[17] Johnson, M. (1987). *The Body in the Mind*. Chicago, IL: University of Chicago Press.

[18] Lakoff, G. and Johnson, M. (1999). *Philosophy in the Flesh: The Embodied Mind and Its Challenge to Western Thought*. New York: Basic Books.

[19] Lakoff, G. and Núñez, R. E. (2000). *Where Mathematics Comes From: How the Embodied Mind Brings Mathematics into Being*. New York: Basic Books.

[20] Rizzolatti, G., Fadiga, L., Fogassi, L., and Gallese, V. (1997). "The space around us". *Science*, 277, 190–191.

[21] Fenyvesi, K., Lehto, S., Brownell, C., Nasiakou, L., Lavicza, Zs., and Kosola, R. (2020). "Learning mathematical concepts as a whole-body experience: Connecting multiple intelligences, creativities and embodiments within the STEAM framework". In P. Burnard and L. Colucci-Gray (eds.), *Why Science and Art Creativities Matter: (Re-)Configuring STEAM for Future-Making Education, Critical Issues in the Future of Learning and Teaching* (vol. 18, pp. 300–336). Leiden: Brill Sense. DOI: 10.1163/9789004421585_018.

[22] Kim, M., Roth, W.-M., and Thom, J. (2011). "Children's gestures and the embodied knowledge of geometry". *International Journal of Science and Mathematics Education*, 9(1), 207–238.

© 2023 World Scientific Publishing Co. Pte. Ltd.
https://doi.org/10.1142/9789811253072_0004

Chapter 3

Mathematical Science Communication for Teachers

Milena Damrau

Faculty of Mathematics/IDM, Bielefeld University,
Bielefeld, Germany
milena.damrau@uni-bielefeld.de

Introduction

One goal of mathematical science communication is to shape people's views of mathematics. It seems that the experiences many people have as children in mathematics classrooms and what they consider as "mathematics" is often different from what mathematicians and/or mathematics communicators have in mind (see, e.g., [1] in which the authors refer to several studies). One reason for this could be how mathematics is commonly taught in school. Studies (e.g., TIMSS video study 1995, see, e.g., [2]) show that mathematics is (or was — the study is from 1995 and studies like this are rare) mainly being taught via routine tasks, and hence as algorithmic procedures. Exploring and problem-solving are rarely present in many classrooms (e.g., in Germany only around 5% and in the US only around 1% of all exercises; [2], p. 199). While algorithms are surely an important part of mathematics and are necessary for our daily lives, being creative, finding strategies to solve mathematical problems, and learning about the diverse mathematical fields might be even more important to develop proper mathematical thinking. If students don't get the chance to experience mathematics in that way in school, it might be difficult to establish views of and beliefs about mathematics that encompass these aspects or change them later on. So why not communicate mathematics in the classroom in a more diverse way, taking into account the various aspects of the science of mathematics, and using different approaches? To implement this, teachers should be considered as multiplying factors and therefore as important math communicators.

This chapter presents the development and conduct of two university courses on mathematical science communication for teachers, which were guided by the following two questions:

1. What can teachers learn from mathematical science communication (projects)?
2. How can those responsible for teacher education optimally equip future teachers for their role as ambassadors for the science of mathematics?

What Can Teachers Learn from Mathematical Science Communication?

This section gives an overview of the similarities between many mathematical science communication projects and suggests practices from which teachers can learn. While mathematical science communication projects are very diverse, a lot of them share similar goals, namely:

- Show the diversity of mathematics (besides school mathematics and algorithmic procedures).
- Raise public awareness of mathematics (as a recent research field).
- Shape people's attitudes toward mathematics.

These goals can (and do) overlap and are of course not comprehensive. But focusing on these three was helpful to me when I started planning my first university course on mathematical science communication for teachers. Changing people's/children's attitudes toward mathematics is probably the most challenging and complex goal, not only because it usually takes time. People's attitudes toward and beliefs about mathematics is a recent and huge research field in mathematics education and it would exceed the scope of this chapter to go into details. But it seems that a positive attitude toward mathematics correlates with higher performance (see for example [3], chapter 10, where attitude was measured through different scales, including "students like learning mathematics", "students value mathematics", and "students confident in mathematics"), even though a causal link couldn't be shown yet. In any case, it is of high interest for mathematics teachers to foster students' interest in mathematics and to enhance a more positive attitude.

Among other factors, Yager and Tamir found that using diverse teaching material and content, collaborative learning in projects, and considering authentic problems play a key role in engaging students with a topic [4]. Showing students the diversity of mathematics (the second goal stated above) can therefore have a positive effect on their interest in and attitude toward mathematics. There are many mathematical science communication projects that provide content to show how diverse mathematics is. To name just a few, the nonprofit organization *IMAGINARY* (see Chapter 8 or [5]) developed computer programs about algebraic geometry, artificial intelligence, and map projections, for example, the mathematics museum *Mathematikum* in Giessen, Germany, has lots of (hands-on) exhibits on different mathematical topics like fractals, probability, symmetry, and optical illusions [6], and the YouTube channel *numberphile* not only provides a huge collection of videos on different topics — many of them also focus on recent mathematical research problems [7]. Therefore, teachers can learn about the diversity and relevance of mathematics by simply getting to know and engaging with these projects. But if we want to influence their teaching, we also have to show and discuss with them how they can use the content from mathematical science communication projects in the classroom.

In the following section, I provide two examples of university courses for mathematics students who want to become teachers. The first one focused more on the teaching *about* mathematical science communication. The second one was more practice-oriented and involved specific ways of using the content of mathematical science communication projects for teaching in the classroom.

University Courses on Mathematical Science Communication for Teachers

The idea I had in mind when developing the first course being outlined in the following section was to show students who want to become teachers how they and their teaching practice can benefit from mathematical science communication (projects). This can be achieved by showing and discussing specific examples and encouraging them to create their own activities.

Due to limitations given by the specifications of the study program, not as many students as hoped for attended this course (more about this in section "Difficulties and Feedback of the Participants").

A first attempt

The first university course on mathematical science communication for teachers I developed has been conducted at Bielefeld University, Germany, in March 2019. It was open for all students who want to become maths teachers, regardless of the school type (e.g., primary and secondary school). The main goals of the course were that students learn:

- About different approaches to understand mathematical content.
- How mathematics can be communicated in a tangible, inclusive and creative way.
- About mathematical content outside of the school curriculum to extend their own knowledge and to be able to enrich their teaching with diverse topics in the future.

The course took place on two consecutive days and a third day a week after. During the first one and a half days, the students learned about what mathematical science communication is, what its main goals are, why it is important, and — most importantly — what kind of mathematical science communication projects exist in Germany, Europe, and the world, respectively. We then discussed what we liked about each of the maths communication projects and why this is the case. The projects we specifically focussed on were:

- The math museum *Mathematikum* in Giessen.
- The nonprofit organization *IMAGINARY*.
- The YouTube channel *numberphile*.

After the students got familiar with the projects and their content, we discussed how these — and mathematical science communication in general — can be useful for mathematics classrooms.

The overall goal was that — based on what the students learned about mathematical science communication during the first half of the course — they all create hands-on exhibits/activities which can be used

Figure 1: First ideas for an Escher tessellation activity using the *Knabbertechnik* (Christina Klaus).

in the mathematics classroom. At the end of the second day, the students discussed ideas for their own projects. I helped them to make these ideas more specific and to find ways for them to implement or build their exhibits. They then had around 1 week to finish their projects.

After 1 week, the students presented what they came up with. One student created an Escher tessellation activity based on the so-called *Knabbertechnik* (from the German word *knabbern* — which means "to nibble" — since one *nibbles* off a part of a polygon; see Figure 1). The target group was children in primary school. Another student wanted to make fractions more tangible for children, so he designed an activity where children could experiment with slices of pizza in different sizes (different denominators). He built a type of pizza oven out of wood to make the experience for the children more fun. One other student made a *GeoGebra*[a] activity using fractions as well. The reason they both chose fractions as the mathematical topic was because they thought it was a particularly difficult concept for the students. Therefore, the students tried to create activities that make fractions more tangible. Overall, it was quite difficult for them to come up with their own ideas since they never had to be creative in that way before. The other students that attended the course did not present their ideas.

---

[a] *GeoGebra* is an interactive geometry, algebra, statistics, and calculus application, see https://geogebra.org.

We ended the course with a visit to an exhibition about stochastics at a local museum. The exhibition is part of a touring exhibition by the *Mathematikum* in Giessen. We engaged with the exhibits and discussed if and how they could be used in the classroom. Since the students analyzed exhibits by the *Mathematikum* on the first day based only through pictures and descriptions, it was interesting for them to really see and engage with the exhibits, and to apply what they have learned. It was therefore a nice way of ending the course.

Difficulties of and feedback from the participants

Unfortunately, the course system at German universities is often not that flexible, which was the biggest problem in organizing and conducting the seminar. Students can only get credit for courses that fit into specific modules they have to cover during their studies, and I wasn't able to convince our faculty that this seminar fits into one of these modules. As a result, students who attended the seminar got no credit apart from a confirmation of participation. Six students (out of over 30 that originally enrolled) attended the course nevertheless.

Despite the small number of participants, I evaluated the seminar. All students had never heard of mathematical science communication and had never been to a math museum before, nor did they know about any of the projects we discussed in the course. Learning about the math museum *Mathematikum* and their exhibits or the YouTube channel *numberphile* alone made the seminar worthwhile for the students. They enjoyed learning about different ways to engage children/people with mathematics, learning about mathematical topics they had never heard of before, and creating their own mathematics activities, which they can later use in their classroom.

The students also recognized the main limitation being the school curriculum in Germany, which is very strict. It is already difficult to teach the prescribed content in a specific amount of time. All students mentioned that they intend to use approaches and content they learned about in this course in the future. However, some of them remained skeptical as to whether they would manage to do so.

A second university course

To ensure that participants got credits for the next course, my colleague Julia Streit-Lehmann and I developed a seminar that fits more directly into one of the mandatory modules. A direct consequence was that we made sure that all the content we chose for the seminar can somehow be linked to the school curriculum.

The seminar took place during the summer semester of 2020 and was named "Mathematical Experiments in the Classroom". Around 90 students enrolled and finished the course. Due to COVID-19, the seminar had to be conducted as an online class. We met every week at the same time via video conference. The overall learning goals were that students:

- Get to know around 10 hands-on exhibits from the *Mathematikum* and learn about mathematical content that is usually not explicitly part of the curriculum (but still relevant and connected to it).
- Identify the mathematical content of the exhibits.
- Get to know possibilities of teaching mathematics in a more tangible way compared to traditional teaching methods.
- Evaluate the potential that exhibits might have for using them in the mathematics classroom.
- Consider ways of using the exhibits for different levels of difficulty.
- Document their learning process by creating a profile (*Steckbrief* in German) for each exhibit.

We did not explicitly discuss mathematical science communication (projects) in this course, but we introduced the students to corresponding content (see Table 1). During the first part of each class, we introduced a new topic. We mainly used exhibits from the *Mathematikum* and presented the content via slides and/or a document camera, but for some topics we showed videos (e.g., by *numberphile*), apps by *IMAGINARY*, introduced the students to *mathigon*, an interactive textbook [8], and added content we produced ourselves. We had mainly two criteria when we chose the content for our seminar: it should have some kind of surprising element to motivate the students to learn more about the topic, and it should have

Table 1:   List of course content.

Topic	Name of *Mathematikum* Exhibit	Additional Materials Used
Galton board & binomial distribution	*Galton Brett*	App *Galton Board*
Exponential growth	*Die Geschichte vom Schachbrett*	—
Fibonacci numbers	*Fibonacci-Zahlen*	*mathigon* website
Birthday paradox	*Zwei an einer Linie*	—
Tower of Hanoi	*Turm von Ionah*	—
Binary numbers	*Binäres Klick-Klack*	Binary card trick
Traveling salesman problem	*Die Deutschlandtour*	A *GeoGebra* activity I created[b]
Map projections	*Kürzeste Wege auf dem Globus*	App *Mappae Mundi* by IMAGINARY and a video
Operative proofs	*Unendlich viele ungerade Zahlen*	—
Unexpected shapes (Möbius strip and cut loops)	*Das Möbiusband*	Videos by *numberphile*

a connection to the school curriculum and therefore direct possibilities for use in the classroom. During the lesson on the traveling salesman problem, we seized the opportunity to address recent mathematical research. I usually try to make students aware of mathematics as a recent research field in many of my courses. But often I have the feeling that it is a bit overwhelming among the actual course content, which for many student teachers is often already challenging.

After the input, the students had to create a one(/two)-page profile for the "topic of the week", for which we provided a template. It included a description of the exhibit, how it is used, what mathematical content it is about, and what other mathematical competencies (e.g., problem solving and/or reasoning) it fosters (the students had to specify the corresponding parts of the school curriculum), what prior knowledge is needed to engage with the exhibit, and how the exhibit can be used in the classroom with different levels of difficulty. At the end of the semester, all students had a collection of 10 profiles for hands-on exhibits they can use for their teaching in the future (see Figure 2 to get an idea of how such a profile looked like).

[b] https://www.geogebra.org/m/mfezqaax.

Gedrehte Acht

Beschreibung:

Für das Experiment werden zwei Papierstreifen benötigt. Diese sollten ungefähr gleich breit und gleich lang sein. Außerdem werden eine Schere und Klebstoff (Klebeband oder Klebestift) verwendet. Die Papierstreifen werden zu einer bestimmten Form zusammengeklebt, die anschließend zerschnitten wird. Die Aufgabe ist es, durch eigene Vorstellungskraft, vor dem Zerschneiden, zu überlegen, welche Form man am Ende erhält.

Durchführung:

Die Papierstreifen werden jeweils zu Kringel zusammengeklebt. Dafür werden ihre Enden mit dem Klebstoff verbunden. Die zwei Kringel werden anschließend aneinandergeklebt. Dabei wird einer der Ringe allerdings um 90° gedreht, bevor er mit dem anderen Ring, der in der Ausgangsposition bleibt, zusammengeklebt wird (siehe Bild). Daher kommt auch der Name „gedrehte Acht". Nun soll überlegt werden, welche Form sich ergibt, wenn beide Ringe an ihrer Mittellinie durchgeschnitten werden. Die Mittellinie ist die Linie, die den gleichen Abstand zu den beiden Außenkanten eines Kringels hat.

Anschließend wird an der Mittellinie der Kringel mit einer Schere entlanggeschnitten. Die Form, die dabei entsteht, ist erstaunlicherweise ein flaches Quadrat.

Inhaltsbezug	Prozessbezug
Grundschule:	**Grundschule:**
Raum und Form:	Problemlösen:
-Raumvorstellung	-Problemstellung erschließen und lösen
-Körper (und ebene Figuren)	-Problemstellung wiedergeben
	Argumentieren:
	-Vermutung aufstellen und begründen
	-Vermutung überprüfen
	Kommunizieren:
	-Gemeinsame Bearbeitung der Aufgabenstellung in Kleingruppen
	-Beschreiben
Sek I:	**Sek I:**
Geometrie:	Argumentieren/Kommunizieren:
-Beschreibung ebener und räumlicher Figuren	-Vermutung aufstellen und begründen
-Charakterisierung von Körpern	-Vermutung überprüfen
-Körper herstellen	-Gemeinsame Bearbeitung der Aufgabenstellung in Kleingruppen
-Anzahl Oberflächen und Kanten bestimmen	-Beschreiben mit mathematischen Fachbegriffen
	Problemlösen:
	-Problemstellung erschließen und lösen
	-Problemstellung wiedergeben
	-Ergebnisse reflektieren
	Werkzeuge:
	-Lineal nutzen (gleich breite Streifen abmessen)

Benötigtes Vorwissen:

-Grundbegriffe: Oberfläche, Kante, Wörter zum Beschreiben räumlicher Figuren

-Übung im Bewegen von Körpern in der Vorstellung

Differenzierungsmöglichkeiten:

Niveau 1:

Die SuS basteln etappenweise die gedrehte Acht:

1. Streifen:
Wie viele Flächen und Kanten hat ein Streifen? Male die Flächen in unterschiedlichen Farben an.
2. Kringel:
Wie viele Flächen und Kanten hat ein Kringel?
3. Gedrehte Acht:
Wie viele Flächen und Kanten hat die gedrehte Acht?
4. Folge-Aufgabe:
Kannst du berechnen, wie viele Flächen und Kanten eine Girlande aus zwei gedrehten Achten hat?
Überlege dir weitere Aufgaben und tausche dich mit einem Partner aus.

Niveau 2:

Die SuS basteln die gedrehte Acht. Wie es das Experiment vorsieht, soll nun die Vermutung aufgestellt werden, welche Figur dabei rauskommt, wenn die Mittellinien aufgeschnitten werden. Die Vermutung soll aufgeschrieben werden. Anschließend soll im Plenum diskutiert werden, welche Vermutung wohl die richtige Lösung ist. Zur Anregung werden Antwortmöglichkeiten gegeben und die SuS sollen sich zu jeder Antwortmöglichkeit äußern.

Anschließend werden die Mittellinien aufgeschnitten. Die Aufgabe lautet nun:

Versuche durch farbige Markierungen auf der Oberfläche oder den Kanten der Streifen zu erklären, wie das Ergebnis zustande kommt. Dafür kannst du immer wieder die gedrehte Acht formen, bis du mit deinem Ergebnis zufrieden bist. Versuche nun einem Partner zu erklären, warum du welche Markierung gemacht hast und wieso aus der gedrehten Acht ein Quadrat geworden ist.

Niveau 3:

Du kannst bereits das Experiment mit der gedrehten Acht. Ergänze die gedrehte Acht um einen weiteren Kringel.

Überlege dir nun, welche Figur dabei rauskommt, wenn du die Figur wie die gedrehte Acht zerschneidest. Bastel das Experiment nach und finde es anschließend heraus. Erkennst du ein fortlaufendes Muster? Wenn ja, welches? Wenn nein, vermute wieso.

Vermute, was aus einer „gedrehten Acht" mit vier Kringel-Elementen beim Zerschneiden der Mittellinie entsteht.

Links: mit 3 Kringel

Rechts: mit 4 Kringel

Figure 2: A profile on unexpected shapes (cut loops) created by my student Mandy Höhne (in German).

Evaluation of the second course

The second course was evaluated by the Faculty of Mathematics via a standardized and anonymous questionnaire. Additionally, the students were asked to answer some specific questions about their experiences in this seminar (anonymous as well). The evaluation was very positive. The students agreed that the ratio between theory and praxis was balanced, which is usually very important for them. In the comments, some students pointed out how experiments could help to engage children with mathematics. Many stated that they liked the structure of the seminar, and that they appreciated access to a rich collection of activities to use for their teaching. One student commented the following:

> *"One of the best if not the best didactic course during my studies. The course was of personal and professional value for me. It was fun to engage with the contents. Especially in the way the materials were presented. Each week there was a WOW moment. I learned a lot in this seminar for future teacher practice and this is something very important for me in a didactic course."*
> [Translated from German by the author]

Since these were some of the main goals of our seminar, it was very satisfying to get such feedback and I hope many students can put what they have learned into practice in the future. The statement also supports our assumption that experiencing "WOW moments" has a positive impact on students' motivations.

Of course, some of the comments were a bit more critical. These referred mainly to two aspects: the amount of course work and the redundancy in writing a profile every week.

Conclusions

One has to be honest: most (future) teachers already have a lot on their plate and during their education they have to cover many other (mandatory) courses. Gaining additional knowledge and experience is therefore sometimes difficult and too time-consuming for them. Furthermore, school and/or university systems often leave little space for outside-the-box activities. One cannot and shouldn't expect teachers to keep up with recent mathematical research, since this is simply beyond their scope of possibilities.

However, many students expressed how important and relevant the content of such courses is, which I completely agree with. I am planning to conduct the seminar about "Mathematical Experiments in the Classroom" again in the near future, maybe with other or additional topics. Since I can relate to the criticism about the redundancy in creating the profiles, I would address this when planning future course iterations.

Teachers can and do learn from mathematical science communication. During the development and implementation of the two university courses, I realized that it is not that important to teach them *about* science communication and it is not enough to just introduce and engage them with these projects. In my experience, teachers learn the most if you select projects and content that is new to them in some ways but still connected to the mathematical school curriculum. In that way, they can benefit from the different methods of communicating mathematics and enrich their teaching with content that motivates them and — hopefully — their students, while simultaneously aligning with the school curriculum. Such alignment allows for feasible use of the content and experimental strategies for learning mathematics, increasing the likelihood of actioning what the student teachers have learned.

References

[1] Loos, A. and Ziegler, G. M. (2015). "Gesellschaftliche Bedeutung der Mathematik". In B. Schmidt-Thieme, R. Bruder, L. Hefendehl-Hebeker, and H.-G. Weigand (eds.), *Handbuch der Mathematikdidaktik* (pp. 3–17). Berlin, Heidelberg: Springer.

[2] Hiebert, J., Stigler, J. W., and Manaster, A. B. (1999). "Mathematical features of lessons in the TIMSS video study". *ZDM — Mathematics Education*, 31, 196–201.

[3] Mullis, I. V. S., Martin, M. O., Foy, P., and Hooper, M. (2016). "*TIMSS 2015 International Results in Mathematics*". TIMSS & PIRLS International Study Center.

[4] Yager, R. E. and Tamir, P. (1993). "STS approach: Reasons, intentions, accomplishments and outcomes". *Science Education*, 77, 637–658.

[5] Website of *IMAGINARY,* https://imaginary.org.

[6] Website of *Mathematikum* in Giessen, https://www.mathematikum.de.

[7] Website of *numberphile,* https://www.youtube.com/user/numberphile.

[8] Website of *mathigon,* https://mathigon.org.

Chapter 4

Math in the City: Designing a Math Trail for High School Students

Francien Bossema*,†,‡, Charlotte Zwetsloot*, and Ionica Smeets*

*Science Communication and Society Department, Leiden University, Leiden, The Netherlands
†Computational Imaging, Centrum Wiskunde & Informatica, Amsterdam, The Netherlands
‡bossema@cwi.nl

This chapter describes the development of a math trail for high school students. In 2016, we developed this trail through Leiden (The Netherlands) during a student project for the Science Communication and Society specialization, a track for master students at the Faculty of Science at Leiden University. Our aim was to provide a guided trail through the city that links everyday sights to mathematical concepts within the curriculum of high school students between 13 and 15 years old. The entire project was carried out in 3 weeks. We did background research, consisting of literature reviews, target audience surveys with school children, and focus groups with teachers. Based on the conclusions from this background research, we developed questions that suited both the goal to make the math trail a fun experience that makes math less abstract and the goal to include questions from across the curriculum. In this chapter, we would like to share our insights from the background research and our experiences in developing a math trail. We moreover aim to provide those who are interested in designing a math trail in their city with a practical step-by-step plan and checklist.

Introduction

There are multiple ways to describe a mathematics trail. For example, Shoaf, Pollak, and Schneider (2004) write: "A mathematics trail is a walk to discover mathematics. (…) The math trail map or guide points to places where walkers formulate, discuss, and solve interesting mathematical problems" [1]. Zender writes: "A mathematics trail is like a sightseeing tour, but the objects are of mathematical interest. One can discover or practice math at the hand of these objects" [2].

Leiden

Math trail

The Morspoort in Leiden

Regardless of slight differences in definition and design, the goal of math trails is similar: to link mathematics to practical situations, and to promote learning and engaging with mathematics in a playful and motivating setting. Math trails are ultimately suitable to engage students in problem-solving, to make connections between concepts and applications, and to apply the skills in a meaningful context [3]. Cross describes the benefits that arise because the activity takes place outside the classroom: the problems are real and relevant, pupils gain confidence by applying their knowledge in practical situations, the discussion that will be stimulated helps them to communicate their knowledge effectively, they need to accurately record their data to be used later in the classroom and they will meet the same ideas and concepts in different contexts [4]. Specifically for a city walk, the assignments should not be solvable without the surroundings and the topics should have been discussed in class before [5]. Mathematics comes to life, because students have to engage both physically and cognitively [6]. Math trails for example exist in The Netherlands[a] and in other countries.[b]

Previous literature on math trails details the development of trails for primary school students [3,6,7], high school students [5], teachers [8,9], a general audience [1], and the development itself as an exercise for the students [6,10]. Zender (2020) describes the development of math trails and research on their effectiveness in the last 40 years [2].

During a 3-week student project in December 2016, the Leiden Math Trail for high school students was developed.[c] The project was part of the Science Communication and Society specialization, a Master Track of the Faculty of Science at Leiden University, The Netherlands.[d] The first two authors and another student worked full-time during these 3 weeks to perform literature reviews and target audience research, supervised by the third author of this chapter. The goal was to develop a route through the city and design questions linked to sights with the appropriate underlying concepts and to combine all this into a booklet to be used by high school children to explore mathematics in the city.

The uniqueness of the student project described in this chapter lies in the integral scientific approach, from background and target audience research to evaluation in a short time period. Zender describes the lack of systematic research on math trails in [2]. The goal of the student project was to have a usable and tangible product based on scientific research and methods and to include as much background research and evaluation as the short time limit of 3 weeks allowed. The background research led to insights that changed the approach of the math trail and was a valuable part

---

[a] For example, in Utrecht: https://hanswisbrun.nl/winkel/wiskundewandeling-utrecht/, and Groningen: https://www.rug.nl/sciencelinx/wiskundeopstraat/ [Accessed on May 4, 2022].

[b] For example, in Cambridge: https://www.cambslearntogether.co.uk/cambridgeshire-school-improvement/cambridgeshire-english-and-maths/cambridgeshire-maths-trails/ resources and https://mathcitymap.eu/en/ [Accessed on May 4, 2022].

[c] The math trail can be downloaded from the Leiden University website here: https://www.universiteitleiden.nl/en/news/2017/01/math-trail-leiden [Accessed on May 4, 2022].

[d] For more information on the specialization programme see: https://www.universiteitleiden.nl/en/science/science-communication-and-society [Accessed on May 4, 2022].

of the development of the project. An evaluation was moreover performed in 2020 for the purpose of making the research presented in this chapter as comprehensive as possible.

In this chapter, we will detail the research steps that were taken to ensure a match between the target audience and the product, elaborate on the designing process, and show the road that led to a successful final product. We will also include a short evaluation with teachers who have requested the teacher's guide in the time since the publication of the trail. The goal of this chapter is to show that in a relatively short period, a math trail can be developed that suits the needs of teachers and high school students, based on scientific ideas and background research. We hope to inspire future designers of math trails with the experiences and insights from this project and to show the added value of research. We will also provide them with a step-by-step plan, checklist and time schedule to make sure their contribution is of value to teachers, students, and of course any other interested people. We believe that our results can be an inspiration for other science communication projects, since it shows how to easily incorporate the target audience in the design of an activity.

Goals

The math trail was developed with the aim to create more enthusiasm for mathematics amongst high school students by engaging them actively in a "treasure hunt"-style walk-through the city. The goal was to actively let them interact with math applications in daily life and show how math is present all around us. One of the underlying goals was to create a fun addition to high school lessons, fitting the curriculum of high school students, and that could be used by teachers as part of their program. The math trail aims to broaden pupils views on mathematics, with the hope to improve the general image of mathematics and the relation pupils have with this subject.

Audience target audience description

The main target audience for the math trail is high school students between the ages of 13 and 15 of the Dutch middle (HAVO) and higher (VWO) level and by extension their math teachers. Although the trail is developed for this main audience, it is freely available and there is therefore a large secondary audience, consisting of anyone who is interested in mathematics and city walks.

Target audience research

To investigate the needs and interests of our main target audience, the teachers and the pupils, audience research consisted of two parts: focus groups with mathematics teachers and surveys with high school children.

Focus groups with high school teachers

We held two focus group discussions of 5–10 minutes in an informal setting with four or five high school teachers. The structure of the focus groups and possible questions can be found in Table 1. Most of the teachers taught the lowest 3 years (12–15) of the middle and higher level of high school.

The teachers in the focus groups agreed unanimously that their pupils would be eager to do a math trail during school hours and that a math trail is well suited to motivate pupils by showing them the applications of mathematics in the outside world. Especially in the context of an active assignment in small groups. One of the teachers remarked: "The

Table 1: Focus group structure.

1. Briefing
   • What is the goal of the focus group?
   • Ask permission for transcribing and possible use of quotes.
2. Open questions and discussion
   • What do you think of when we say "math trail"?
   • What would you like to be included in a math trail?
   • What are the conditions and requirements to do a math trail with your students?
   • Do you think the students would be interested in a math trail (during school hours)?
3. Survey: Small sheet of paper to get personal information
   • Name and email (optional for sending results and trail).
   • School and city.
   • Level of high school and grades of the taught classes.
   • Optional if a publication is foreseen: Do you wish your name to appear in acknowledgments of a possible publication?
4. Debrief
   • Thank them for their contribution and if they wrote down an email address they will receive the trail when finished.
   • Hand participants a small note with contact details.

goal would be that the students would get more context to the mathematics, which they otherwise only see in their textbook. I would like them to see it in real life, to make them experience it more."

The teachers agreed that the trail questions needed to have enough depth. The "puzzle" part of the trail needed to be challenging. One of the teachers said: "It is not a problem if they have to spend 10 minutes on a question". The questions should have more content than just calculations, but a one-to-one overlap with the curriculum was not a must. The teachers viewed the trail as an addition and extension of the already discussed theory. There should be a lot of variety in the questions and it would be an asset if the most interesting sights of the city were featured in the trail.

Most of the teachers would link the math trail to an assignment, add a competition element or give a mark for it. They expected that the students would not be motivated enough if there was no obligation attached. They estimated that the students would be able to concentrate on a trail of 1.5–2 hours maximum. The questions should have sufficient space in between them to allow groups of students to depart one after the other without all accumulating at the same question. All the teachers would be interested in the trail but saw different opportunities. One would like it to fit in one lesson (or a double lesson), for example, to complete in the last lessons before the holiday. Another thought of walking the trail in a project week, so that the students would have enough time for an extensive trail. One other teacher had the idea to make designing a trail for the 13–14-year-olds a project for the older students (15–16).

Surveys with high school students

We conducted surveys with high school children in the second and third grade (13–15-year-olds) of the higher-level high schools. The survey questions can be found in Table 2. 97 students filled out the survey in total, with 47 in the second grade and 50 in the third grade.

The results of the survey showed clearly that most of the students would not do a math trail in their free time (92%) but would be interested in a math trail during school hours (95%). Most of the students had a neutral opinion about mathematics (55/97), 31 out of 97 said they liked

Table 2:   Questions for the target audience survey.

1. Short introduction: Why this survey?
2. Math and math trails
   - What is your opinion on mathematics? Likert Scale, five choices from "I don't like it at all" to "I like it a lot". Give a neutral option.
   - What do you think of when we say "math trail"?
   - What would you like to do in a math trial through the city?
   - Would you like to do a math trail in your spare time? Yes/No.
   - Would you like to do a math trial instead of a regular lesson? Yes/No.
3. Demographic questions
   - What is your gender? Girl/Boy.
   - What is your age?
   - What school are you attending?
   - What is the level of the school?
   - What grade are you in?
4. Example questions (note, don't answer them, just mark the one you like best!).
   - We gave two options of questions and asked the students to pick the question that appealed to them most. We did this for two topics: geometry and combinatorics.

mathematics and only 10 did not like mathematics at all. Most of the students had quite an accurate idea of what a math trail would be; they expect it to be a walk during which exercises have to be done.

To find out their preference for a certain type of question, we included two example questions: one about calculating angles and one about combinatorics. For the question about angles, there were two options: a passive question involving a clock, which was only solvable by thinking, and an active question involving a lantern, which was only solvable by walking around. A slight majority of the students preferred the passive question (56%). The goal of the combinatorics question was to assess whether the pupils would prefer a question that was directly linked to a site over a question that could be answered anywhere. The first question was one about the possible combinations of the numbers in a pincode, the second a question about the possible ways to color the coat of arms of the city Leiden. The underlying idea was that the first could be done in any city, whereas the second was specific to Leiden. The pupils were in favor of the pincode question (75%). This disconfirmed our hypothesis that a link with the city would be preferred. We suspect that this can be partly attributed to the fact that the question on the coat of arms was much

Table 3: Checklist for the target audience survey.

- Ask for demographic details.
- Ask for their likes and dislikes.
- Be explicit in what you want them to do.
- Make sure example questions are similar, do they test what you want to test?
- If possible, do a test run with a few pupils.
- Try to get a varied audience, to avoid bias toward certain students or levels.

lengthier due to the historical background being included. We therefore think that this question may have tested whether students like longer or shorter questions, which was not the goal of this question. To assist future math trail developers, our experiences are summarized in a checklist for target audience research in Table 3.

Many pupils gave non-math-related answers to the question "What would you like to do in a math trial through the city?", such as "not being in school", "freedom", "eating during the walk". For the survey that took place after the first, we therefore added that we would like to hear any ideas they had on questions that we could use. Some noteworthy answers were: calculate the height of a church, binary puzzles, calculate how much can you buy in a shop when you've got 10 Euros, calculate how much water there is in the canals, logical thinking.

Conclusions from the target audience research

Based on the target audience research, we concluded that the most important goal of a math trail to us is to create enthusiasm amongst the students and to bring them in contact with mathematics in real life, which can make math topics less abstract. An engaging activity in the open air gives the students a positive attitude toward mathematics, which is then hopefully transferred to the classroom. To achieve this, it can be useful if they are actively working on mathematics in small groups, independently from the teacher, to promote self-guided learning. A competition element or marked assignment attached to the trail could encourage the pupils to take the trail seriously. The surveys clearly showed that the students would not do the trail in their spare time but were very interested in it during school hours. Teachers are enthusiastic about the concept of a math trail, and they would like to see varying and challenging questions. It is moreover important to link the questions to sights in the city and have enough space between questions. Lastly, the trail should take 1.5–2 hours.

Format and development

Target audience

The mathematician who suggested the idea for a math trail as a science communication student project, originally wanted to create a math trail for 15–16-year-olds, to do with their families in their pastime. We suspected that the students would not be overly enthusiastic to do extra math exercises in their pastime and our surveys confirmed this. The teachers of the lower grades expected their students to be motivated to do a math trail, during school hours and in groups. We therefore adjusted the target audience to 13–15-year-olds during school hours.

As the trail targets different grades and levels of mathematics, we tried to avoid the need for too much known theory. The questions should be answerable with a basic knowledge of mathematics and logical thinking. Where any extra theory was necessary, we added this theory to the description of the question, without giving the answer away. An example was providing the formula for the volume of a sphere. This also enables a much wider group outside the target audience to do the trail: families, students, and math aficionados.

Design process

The teachers in our focus groups gave a time limit of 1.5–2 hours for a regular lesson, but others indicated that the trail could be longer for a project week or excursion. To accommodate both, we decided to design a trail of about 2 hours with an optional extension, which takes another 45 minutes.

The design process of the walk consisted of a number of steps (see also [3]):

1. Exploration and draft
2. Write route directions
3. Formulate questions
4. Design and check

Exploration and draft

To decide on the route, we started by collecting sights in the city, which the trail should pass. The criteria for including sights was, for example, the historical value of the sight, an aesthetic value such as a beautiful house, the diversity of the trail (streets and a park), and inspiration for questions. Examples of this inspiration are a church, the coat of arms of Leiden (Figure 1 left), the Weigh House, and a beautiful house with orange stones (Figure 1 right).

We were moreover in contact with the local science museum Rijksmuseum Boerhaave, who had at that time a mathematics exposition and were interested in hosting the booklets of the trail, and providing those who walked the trail with the answer sheet at the end. This provided us with a good starting and end point for our trail, as the museum is centrally located in the city and within walking distance from the train station. It enhanced the possibility for an excursion, as the trail could be combined with a visit to the museum and also a good possibility for our secondary target audience to get hold of the trail and only receive the answers at the end. It was moreover possible to find a circular route starting at Rijksmuseum Boerhaave, with a shortcut after 2 hours to design the route plus extension route. During the exploration walks, we took many pictures to record possible question locations and noted any locations that should be included because of the historical value or inspiration for questions. We kept in mind that the route should be quiet enough, but not too secluded, and safe for the students to walk. Streets without cars had our preference

Figure 1: Inspiring sights in the city. Left: The coat of arms of Leiden on the side of the Hooglandse church, right: "The orange house".

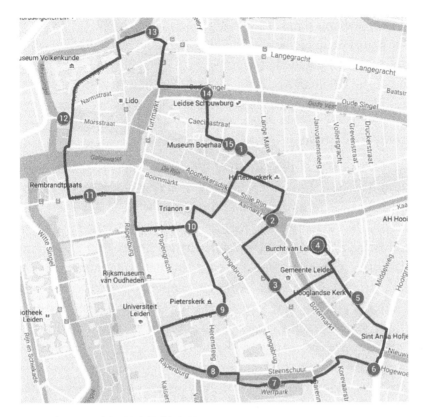

Figure 2:   The route of the Math Trail in Leiden.

over streets with cars. Besides the math element, a trail is more of an outing if there are nice shops to peek into. Based on these criteria and our notes, we used a map to draw a route (see Figure 2). We moreover wrote down our draft questions and noted where any other questions should be inserted to make the time between stops approximately the same.

Write route directions

The next step in the design process was to write clear directions for the chosen route. Two of the three students involved in the design process wrote the directions while walking the trail again. The third then used these directions to walk the trail and check whether the directions were clear. Next, we asked a few friends to walk the route alone or with one of us to identify any mistakes or unclear directions. In total, the route was proof-walked four times. This is an important step in the design process, as the students should be able to follow the route by themselves without a teacher. When you design your own trail, keep in mind that over time some of the landmarks on your route description may change and you have to walk the route now and then to keep it up to date.

Formulate questions

An important goal of the math trail is to make students enthusiastic for math and to show them the applications of math in daily life. Our aim was to keep the questions original and surprising, while also keeping a tight link with the city. Therefore, we have developed most of the questions ourselves. Some questions were inspired by an example from another math trail or an exercise in a textbook. The mathematics in the questions does not always correspond one-to-one to the curriculum, so as to disconnect from the textbooks and use any inspiration from city highlights. We decided, however, to design some questions based on the curriculum, to facilitate the discussion of certain theories by the teachers prior to or after walking the trail. While we extended the questions, we simultaneously wrote the answer sheet. We sometimes included short explanations to make them understandable for the interested people who would do this trail on their own, to meet the needs of the secondary audience.

Diversity in the questions is of importance; this was confirmed by both the background research and the literature. We have taken this into account in several ways. First of all, the questions are diverse in the underlying mathematical concepts. We have noticed in other trails that it was a pitfall to use mainly geometry questions in a trail. Geometry is the most visible in the surroundings, but it is important to have variety in the question topics [9]. One can think of, for example, statistics, probability, modeling, functions, derivatives, variables, and equations. Inspiration can be found in other math trails or in literature.

The second type of diversity can be found in the type of questions: open questions, multiple choice questions, calculations. Also, the underlying goal of a question can be different: application, analysis, and making connections between concepts.

Thirdly, there is a variety in subjects outside mathematics. There is an overlap in mathematics-related and physics-related questions, because many applications of mathematics are in physics. This overlap helps to put mathematics in perspective and combats its image of a subject that stands on its own. For example, we used an existing "wall formula"[e] to discuss Snell's law. In addition, history can be included in a math trail through a city, which may enhance interest in the subject.

As a last consideration, we have based our questions on the learning goals as published by the committee Toekomst Wiskunde Onderwijs (Future of Mathematics Education) [11]: problem-solving and analytical thinking, modeling and algebra, ordering and structuring, manipulation of formulas, abstraction, logical reasoning and proofs. We have ensured that all these concepts are present in at least one of the questions in the trail. In Table 4, we point out which concepts are present in each question of the trail.

Table 4: Learning goals per question.

	Problem-solving and analytical thinking	Modeling and algebra	Ordering and structuring	Manipulation of formulas	Abstraction	Logical reasoning and proofs
1. The Lottery	X	X				X
2. The Weigh House	X	X				X
3. The Rijnland foot					X	X
4. The fortress		X		X	X	
5. Leiden's coat of arms	X	X				X
6. Snell's law				X		
7. The gunpowder disaster				X		X
8. The orange house	X		X			

(Continued)

[e] https://muurformules.nl/?ln=en [Accessed on May 4, 2022].

Table 4: *(Continued)*

	Problem-solving and analytical thinking	Modeling and algebra	Ordering and structuring	Manipulation of formulas	Abstraction	Logical reasoning and proofs
9. The Pieterskerk	X					X
10. The lantern		X	X		X	
11. The stepped gable		X	X			X
12. The cannon				X	X	
13. The windmill					X	X
14. Gables			X			
Bonus: The bridge problem	X	X	X		X	X

Design and check

For the final design, we decided to choose the simple format of a booklet, answer sheet, and teachers' guide. One reason for this was that it made it possible to meet our time limit of 3 weeks. We moreover reasoned that this was the easiest way to make the trail available to a wider public, as it could be made available both through a free PDF on the website of the university and in printed form together with the answer sheet at Rijksmuseum Boerhaave. The final version of the math trail was designed using InDesign.

The booklet consists of a map with the route and on each page a question followed by the direction to the next stop. There is room for the answers on each page, there are some empty pages at the end to use as scrap paper, and there is a printed set square tool on the last page, so that the users of the trail only need to bring a pen.

As we decided to include historical sights in our route, we wanted to include some information about them for those who are interested. As this information is not related to the mathematics in the trail, we put the information into frames on the side, to make it clear that it was additional information and could be easily ignored by those who were not interested in this extra context.

The main audience was high school students who would do this trail with the school, therefore we extended the answer sheet to a teachers' guide. In this guide, extra practical information was included, such as the presence of a small café in the park in the middle of the route, where the teacher could take post to meet the pupils halfway. Another tip was that the students could bring their lunch and eat that in the park for a break along the way. The teachers' guide also expands the answer sheet with suggestions for continuing the discussion about the questions in the lessons, using the trail as an original method to start a topic from the theory to be taught. The teachers' guide is available via email at the university upon request of the teacher.

The last step in the development of the math trail is to thoroughly check all the material and to ask for feedback from people who were not involved in the development process to proofread all the material, including if possible, high school teachers. Our trail was developed in Dutch and due to its success, it was afterwards translated to English by the Academic Language Center of Leiden University.[f]

Evaluation

When the trail was just published, the local media showed interest. Three radio programs featured an interview with one of the developers,[g,h,i]

---

[f] https://www.universiteitleiden.nl/en/language-centre [Accessed on May 4, 2022].

[g] https://www.nporadio1.nl/fragmenten/nieuws-en-co/8beb7fab-e27c-45c9-bcdc-106f1ab2e1c1/2017-02-15-wiskunde-in-de-leidse-buitenlucht [Accessed on May 4, 2022].

[h] https://sleutelstad.nl/2017/01/25/nationalisme-pijnstillende-cannabis-en-knuffelhormoon/ [Accessed on May 4, 2022].

[i] Omroep West (local Dutch radio station), program 'Mogge Michiel', January 20, 2017.

it was featured on news and math websites,[j,k] and the local newspaper Leids Dagblad wrote a short article on the trail.[l] It was also published on the university website[m] and announced on the website of Rijksmuseum Boerhaave.[n] Directly after the publication of the trail, there was no evaluation research, as this was not part of the student project and, as it so often goes, we all started working on other projects. However, we have seen people walk the trail in the city and the booklets at Rijksmuseum Boerhaave had to be reprinted half a year later (the first edition consisted of 350 booklets), which shows that there is a considerable secondary audience.

Since the trail was published in 2016, over 40 teachers have requested the teachers' guide. In 2020, when this chapter was written, we sent out a survey to them to evaluate the goals of the math trail. In total 19 out of the 47 addressed persons filled in the survey. Of these, 11 did the math trail with their students and indicated that they would like to do so again in later years. The reactions were predominantly positive. Many teachers did the trail as part of an excursion to Leiden in a project week, combined with a visit to one of the local museums for a different course than mathematics. The number of students for each teacher ranged from 20 to 125, mostly 50 or more per group. There was no grade attached to the trail, but the teachers thought of fun ways of encouraging their students, such as including the unraveling of a code through the exercises with which they could open a pot with candy at a stop with teachers in the middle of the walk. Another teacher included the option of buying "hints" for a reduction in points, to enhance the feeling of competition. The opinion on the level of the questions differed, many mentioned the level to be fine, some too easy, some too hard. Of course, this also depended on the age and grade of

---

[j] https://www.omroepwest.nl/nieuws/3336446/Leiden-heeft-eigen-wiskundewandeling [Accessed on May 4, 2022].

[k] https://pyth.eu/wandelen-in-leiden [Accessed on May 4, 2022] and https://pyth.eu/de-pythagoras-wandeling [Accessed on May 4, 2022].

[l] https://www.leidschdagblad.nl/cnt/dmf20180926_61900602/leiden-is-weer-een-wandelroute-rijker?utm_source=google&utm_medium=organic [Accessed on May 4, 2022].

[m] https://www.universiteitleiden.nl/nieuws/2017/01/voor-de-liefhebbers-van-wiskunde-de-leidse-wiskundewandeling [Accessed on May 4, 2022]; in English: https://www.universiteitleiden.nl/en/news/2017/01/math-trail-leiden [Accessed on May 4, 2022].

[n] https://rijksmuseumboerhaave.nl/te-zien-te-doen/spelen-met-wiskunde-abc/ [Accessed on May 4, 2022].

the students, as these varied from the target group to older students. On average the teachers thought the students enjoyed the activity.

The evaluation clearly showed that many teachers were inspired by the trail. Some modified it, used certain questions for a shorter version of the route, or added extra assignments. One teacher extended the excursion with an assignment afterwards, for which the students had to design some math trail questions themselves and present them to each other and parents for a grade. All teachers thought the trail had definitely added value, because it was "different from other school activities, nice to be outside", "provided the insight that math is present in the world around us" and "showed the practical application of math". One question was mentioned often as too hard because of the practical issue that the students had to stand on exactly the right spot in the street to solve it. There were a few other useful remarks and tips. It was mentioned that at least one accompanying teacher should be a math teacher. To combine with an excursion to a museum, it might be useful to break the walk into even smaller trails than the two provided. This would also solve the issue that for lower grade students a shorter trail was preferred and that it is quite a challenge to guide the number of students (40 or more) along the same trail. One teacher mentioned it was useful for them to walk the trail themselves first, to assess the level of the questions and estimate how much time students would need.

Discussion and Conclusion

In this chapter, we have described the development process of a successful mathematics trail in a 3-week student project. We have shown the added value of the target audience and the background research, and hope that this will inspire others to develop a much-used math trail. For the target audience research, it would have been good to have a wider spread amongst the different levels of education in the target audience. As the surveys in our research have only been filled by students in the higher educational level of high school (VWO), the results may be biased. It is moreover good to do a test run, to make sure one is testing what one wants to test and not, for example, length of the question instead of the topic.

To further the development of new math trails using the experience of this project, time schedules and checklists can be found in Tables 5 and 6.

It would be of great value to do a few test walks with students and teachers of the target group to assess the level of the questions and to find out which type of questions create the most enthusiasm. Moreover, evaluation is recommended to further increase the usefulness of the trail to teachers and students. Besides contacting teachers, for example, diagnostic interviews with students might be of interest if time allows [5]. In order to make sure the trail is of value over the years, it is advisable to walk it yourself every now and then to check the route and directions. The evaluation research has moreover shown that teachers gladly use the trail as input and in this

Table 5: Time schedule and step-by-step plan.

Week 1

Preparation, exploration, and draft
- Design target audience research (i.e., surveys and focus groups).
- Contact the target audience to hold surveys, etc.
- Literature search.
- Perform target audience research and start analysis of results.
- Route draft: walk around the city and take many pictures, keeping track of any inspiration that could be developed into a question and other useful information.

Week 2

Research analysis, route directions, and questions
- Work out literature research and target audience research.
- Write down the route description and draw a map with the stops.
- Ask for and incorporate feedback on the route description.
- Draft questions and keep track of the choices you make.
- Make the visual additions to questions (pictures and graphs).
- Ask for and incorporate feedback on the questions (can be combined with the route description feedback).
- Make a first draft of the trail booklet design.

Week 3

Design and check
- Proof-walk the route and incorporate any improvements and extra inspiration.
- Gather any other information that should be included, such as practical additions (set square and scratch paper) and historical information on the city highlights.
- Finish the questions and visual additions.
- Write the answers to make an answer sheet and optional teacher's guide.
- Layout the booklet (for example in InDesign or another design programme).
- Send the end product to the people who were involved in the process (teachers in the focus group, proof-walkers, feedback-givers).
- If required or desired, write research and development report.

Table 6: Checklist math trail.

Route
- Do you have a clear map, on which the stops are indicated?
- Is it circular and about 2 hours (or with an extension)?
- Is your route foolproof? Can anyone follow the route description?
- Does it include varied scenery: historical sights, a park, a church?

Questions
- Do the questions target the different math areas: problem-solving and analytical thinking, modeling and algebra, ordering and structuring, manipulation of formulas, abstraction, and logical reasoning and proofs?
- Are the questions linked to visual representations?
- Are the questions fun and linked to things that can be seen during the trail?

Design
- Does the design contain: scratch space, a set square, references if needed, information about contact, and how to get the answer sheet?
- Do you have a clear distinction between questions and route description?
- Did you include contact details for requesting the answers and for letting you know when there was a problem following the route as described?

---

format can easily adjust it to their needs and insights. The development of the math trail was an interesting and successful student project and is, 4 years later, still put to good use.

Acknowledgments

We would like to thank the third student who was involved in developing the math trail, Suzanne Kappetein, for her contributions to the project. We thank Peter Kop from the Leiden University Graduate School of Teaching (ICLON) for proposing this project and his supervision during the development in 2016. We thank teacher Roland Lucas, Stedelijk Gymnasium Leiden, for his useful feedback on the trail and the opportunity of engaging his students in our target audience survey. We would like to thank Mieke Thijsseling, Willem Lodewijk Gymnasium Groningen, for distributing the survey amongst her students and filling out the survey after walking the trail with her students. We would like to thank the other teachers and math trail enthusiasts who filled in our evaluation survey, amongst others: Anneke Korevaar, K. Dolderman, Marianne Laponder, N. Dijkstra, Epi van Winsen, and Hanneke Klarenbeek-den Heijer. We thank the Nederlandse Vereniging voor Wiskundeleraren (NVvW) for funding the printing of the booklets and Rijksmuseum Boerhaave for distributing them. We would like to thank vice-dean Han de Winde for facilitating and funding the translation of the booklet to English, and the translator Leston Buell for the translation. Furthermore, we would like to thank the following people: the organization of the Noordhoff Uitgevers Wiskundecongres, all students who have filled out our survey, the teachers who have participated in our focus groups, the people who proof-walked the route and provided their feedback, and Jos van den Broek for providing feedback on the layout of the booklet.

References

[1] Shoaf, M. M., Pollak, H., and Schneider, J. (2004). *Math Trails*. COMAP Inc. ISBN: 0-912843-76-4.

[2] Zender, J. (2020). On the history of mathematics trails. In *Conference Paper for the Sixth International Conference on the History of Mathematics Education*, Marseille, 2020.

[3] Richardson, K. M. (2004). "Designing math trails for the elementary school". *Teaching Children Mathematics, 11*, 8–14.

[4] Cross, R. (1997). "Developing maths trails". *Mathematics Teaching, 158*, 38–39.

[5] Buchholtz, N. (2017). "How teachers can promote mathematising by means of mathematical city walks". In G. Stillman, W. Blum, and G. Kaiser (eds.), *Mathematical Modelling and Applications. International Perspectives on the Teaching and Learning of Mathematical Modelling*. Springer, Cham.

[6] English, L. D., Humble, S., Barnes, V. E., and Trailblazers (2010). *Teaching Children Mathematics, 16*(7).

[7] Smith, K. H. and Fuentes, S. Q. (2012). "A mathematics and ScienceTrail". *APMC, 17*(2).

[8] Druken, B. and Frazin, S. (2018). "Modeling with math trails". *Ohio Journal of School Mathematics, 79*, 43–53.

[9] Barbosa, A. and Vale, I. (2016). "Math trails: Meaningful mathematics outside the classroom with pre-service teachers". *Journal of the European Teacher Education Network, 11*, 63–72.

[10] Hitting the math trail, n.d. https://www.educationworld.com/a_curr/curr403.shtml [Accessed on May 4, 2022].

[11] SLO. (2014). Onderwijzen van wiskundige denkactiviteiten, *Implementatie examenprogramma's havo-vwo 2015*. Enschede.

© 2023 World Scientific Publishing Co. Pte. Ltd.
https://doi.org/10.1142/9789811253072_0006

Chapter 5

New York's National Museum of Mathematics

Glen Whitney

gwhitney@post.harvard.edu

On the appropriately symmetric date of 12/12/12, over three hundred people gathered at the north end of Manhattan's Madison Square Park to celebrate the flipping of a giant switch from 0 to 1. That switch simultaneously turned on the power to a new museum and updated the count of North American museums devoted specifically to the public communication of mathematical sciences. How did this day come to pass?

Conception and Formation of MoMath

In August 2008, roughly a dozen volunteers gathered to ponder the creation of a hands-on museum devoted to mathematics, starting with very little except their own convictions that it was a worthwhile effort. This group consisted of people from a variety of backgrounds: area schoolteachers, board members of an earlier museum effort (see more on that project below), an active parent concerned about the state of math education, and a university professor with a dual career as a mathematical sculptor. Dubbing themselves the "Working Group", they saw an immediate need to proceed on many tracks at once. Some of those efforts, such as the creation of a not-for-profit legal entity, identification and persuasion of Trustees to govern that entity, and general fundraising to finance the project, don't pertain to the concerns of this *handbook*. Articulating and honing the mission and strategies of the organization, selecting what to do and how to do it, and reflecting on insights gained from the group's experiences, on the other hand, constitute core issues in communicating mathematics.

Initially, the activity of the Working Group amounted to little more than monthly meetings. Not knowing where else to start, they focused on the name of the new institution, its purpose, and how to capture that purpose in a written "mission statement". Writing by committee is notoriously awkward, and several months elapsed before the Working Group adopted phrasing that the National Museum of Mathematics (MoMath) uses to date:

> Mathematics illuminates the patterns that abound in our world. The [museum] strives to enhance public understanding and perception of mathematics. Its dynamic exhibits and programs stimulate inquiry, spark curiosity, and reveal the wonders of mathematics. The [museum]'s activities lead a broad and diverse audience to understand the evolving, creative, human, and aesthetic nature of mathematics.

As for the name of the new institution, the Working Group sifted dozens of possibilities, from "Math Museum" to "Mathport" to "Infinity Gallery". Studies indicating that the word "museum" could connote stodginess and/or discourage visitation influenced the ensuing selection process. (See [1] for an example of more recent work along similar lines, in the context of diversity in science center visitation.) Thus, the group initially chose "Math Factory", hoping that the word "factory" would convey the activity and dynamism highlighted in the newly-adopted mission statement.

In parallel, the Working Group cast about for concrete actions it could take to promote the idea of the Math Factory, to attract additional participation in the project, and to begin the program work of the mission. To this end, members of the group began to discuss the nascent institution with friends, acquaintances, and connections, new and old, in both formal and casual settings. By December, this networking led to contact with the World Science Festival (WSF), a new public program in New York City that was planning its second annual presentation the upcoming summer. The WSF had encountered difficulties in identifying and including mathematical content in its debut, leaving possible space for new math-related public outreach. Although the WSF made no hard commitments at this point, the Working Group decided to pursue the opportunity aggressively; the prospect of thousands of visitors at a high-profile event in Manhattan seemed an ideal springboard for the Math Factory.

Of course, this decision begged the question of *what* to present. The Math Factory needed something compelling enough to cement the Festival's interest in providing space to an unproven organization, something of a quality high enough to impress Festival organizers and visitors alike, and something that would help galvanize support for its longer-term goals. The Working Group also felt that it was more effective to *show* people an unfamiliar concept (since "interactive math exhibit" would surely be a novel idea to most supporters), than to merely *tell* them about it. Further, some members of the group had started visiting other science and technology centers and hands-on museums to find inspiration and gather advice, and noted a lack of existing resources for such institutions to present mathematics in an engaging way. Finally, if the Working Group was going to put significant effort into the WSF, it ought to end up with an asset of ongoing value. Therefore, in mid-January, the Math Factory circulated a Request for Proposals (RFP) for the design and fabrication of an approximately 125 m^2 traveling exhibition of physically participatory math activities.

At this juncture, the Math Factory enjoyed a confluence of good fortune. Its fundraising efforts had piqued the interest of a donor willing to sponsor the traveling exhibition up to a half-million-dollar budget. Although the Working Group didn't realize it at the time, the mortgage crisis and subsequent financial distress had created a near-vacuum of museum exhibition projects, leading to a robust response to its RFP at feasible costs. And perhaps most importantly, nobody in the group was experienced enough to know that it's "impossible" to produce a traveling exhibition in only 5 months.

Needless to say, that short timetable dramatically increased the level of activity of the Working Group. It was able to engage Ralph Appelbaum Associates, a well-known museum design firm in Manhattan, and the creative process began. There was an immediate, pressing need for specific exhibit descriptions and specifications. Fortunately, the Working Group had already begun a modest crowd sourcing website where mathematical colleagues and connections of the group were invited to submit exhibit ideas. The team mined this resource and brainstormed as many more concepts as time would allow. The process led to a few dozen possibilities, which were winnowed primarily by seat-of-the-pants intuition of the group and the feasibility of producing design drawings for the concepts identified (which in some cases the design team had only tentative ideas as to how they might be realized).

The most striking example of an exhibition brought to life by this speculative enthusiasm was the square-wheeled tricycle, which would become the iconic centerpiece of the exhibition. Several members of the group knew that Stan Wagon of Macalester College had built a square-wheeled vehicle on a straight track in 1995 [2], but resetting such a vehicle after each trip down the track would be operationally prohibitive for a public exhibition. A circular track would eliminate this difficulty. Unfortunately, no one could see a way to actually build such an exhibit, and the group was on the verge of discarding the concept. However, a single enterprising member (the mathematical sculptor mentioned above) blurted out that it "should simply be a matter of engineering", so the tricycle was put into the fabrication package anyway. And indeed, one of the fabricators, Universal Services Associates, submitted a quote for the device, devising a mechanism that would eventually result in a patent [3].

Other notable components of the exhibition included: a "function machine" with selectable operations that would transform input "number tickets" into their function values; a magnetic wall with hundreds of tiles of various tessellating shapes for free play; walk-on mazes with mathematical twists; a plaza on which visitors could physically attempt to solve the traveling salesman problem; a track that visitors could reconfigure arbitrarily to find the shape (the "brachistochrone") that would make a model car travel the fastest; and an apparatus that used a plane of laser light to highlight cross sections of transparent solids. Since these exhibits had been chosen rapidly, and primarily according to their attractiveness for exploratory play and their feasibility to build, the exhibition lacked an organizing principle or theme. Thus, it was dubbed the *Math Midway* — an eclectic, carnival-like collection of mathematical curiosities and attractions. With a tremendous push by the Working Group, design firm, and fabrication firms alike, the traveling show (which had grown during development to over 400 m^2) opened June 14, 2009 at the WSF, introducing the new math museum to the public for the first time.

Prior Hands-On Mathematics in North America

To obtain a deeper understanding of how the Math Factory reached its public debut and how it chose what to present, it's helpful to look at prior opportunities for public audiences to explore mathematical ideas in an engaging way. The seminal effort in this direction was the 1961 exhibition *Mathematica: A World of Numbers and Beyond*, designed by Ray and Charles Eames.

In that year, the California Museum of Science and Industry opened a new science wing, and turned to the IBM Corporation for support. IBM solicited exhibition ideas from the Eameses, who had already served as design consultants for the firm. The Eames organization apparently took to the assignment with relish, submitting a proposal for a graphically rich collection of static and kinetic displays to show the "joy and excitement that mathematicians find in pursuing their science" [4]. IBM commissioned the exhibition, and *Mathematica* was born.

Since this exhibition debuted early in the history of interactive math and science (predating the hands-on Museum of Science and Technology in Tel Aviv by three years and the Exploratorium in San Francisco by eight) and introduced many novel and influential ideas, it's worth recapping its central elements in some detail. The exhibition features a quirky selection of components: Probability, Celestial Mechanics, Topology, Multiplication, Projective Geometry, and Minimal Surfaces [5]. Each section has as its centerpiece one of the main kinetic and/or participatory displays, so it's easy to suspect that the selection and design of these displays drove the choice of topics, rather than the other way around.

Perhaps the most iconic (and the most popular, per the Constantine *et al.* evaluation [4] of the exhibition for the Boston Museum of Science) piece in the exhibition is the Probability Machine. This type of machine has become reasonably commonplace, featuring as the "Plinko Board" of the long-running American game show *The Price is Right*, and sometimes known as a "Galton Board" (and indeed, available for purchase online in 2020 as a desktop model for under $35 [6]). In 1961, however, it was unparalleled as a museum piece: thousands of balls in constant noisy motion, bouncing left or right at each of hundreds of pins to recreate, over and over, physical approximations to a normal distribution. Although not strictly interactive in the modern sense, the Probability Machine very effectively put visitors on notice that *Mathematica* was a different sort of exhibition, and drew them in to its subject matter that similarly was not usual museum fare.

In fact, all of the section centerpieces are noteworthy in their own right. Celestial Mechanics featured a gravity well simulation funnel in which ball bearings rolled in paths approximating planetary orbits. Although such funnels are now familiar, often as "spiral wishing wells" for coins, Mirenberg's study [7] lists no earlier physically-realized instance. The Topology section included a model train perpetually running on a track along a giant Möbius strip, with its orientation reversed each time it passed the same location. The Multiplication component displayed one of the world's first electronic interactives: when a visitor pressed three numbered buttons, a three-dimensional grid of lights would illuminate, illustrating their product. Only Projective Geometry's centerpiece had no kinetic component, but it did demand audience participation. It consisted of tableaux that appeared very familiar and consistent when viewed through a well-situated monocular eyepiece, but utterly incoherent when viewed from an ordinary side perspective. Finally, a display of unusual soap films anchored the Minimal Surfaces section: at the push of a button, several frameworks (including a trefoil knot and a cube) were lowered into a soap solution and extracted again, leaving behind (with some luck) such bubble configurations as a Möbius strip or a cube-within-a-cube.

Mathematica has enjoyed an almost-unparalleled enduring popularity. It ran for 37 years in its original installation and spawned two additional copies. Among these, a partial list of venues includes the California Museum of Science and Industry, the Chicago Museum of Science and Industry, the New York World's Fair, and the San Francisco Exploratorium. All three copies remain on active display in 2020, despite seeming flaws such as outdated information and culturally biased topic selections, at the New York Hall of Science, the Boston Museum of Science, and the Henry Ford Museum of American Innovation.

At the opposite end of the spectrum, in terms of sophistication of design and budgetary resources, lay the Goudreau Museum of Mathematics in Art and Science, on Long Island, NY. Bernard Goudreau taught high school math beginning in 1965, and became locally renowned for the variety and quantity of mathematical models, games, and puzzles that he and his students would create for his classes. His colleagues felt that these materials could serve, and should reach, a broader audience, and a coalition came together in 1980 to form a small museum with a collection of these items as its seed.

The resulting museum was housed in two former junior-high classrooms in a local community center (under 200 m²). These rooms were stuffed almost to bursting with a hodgepodge of geometric models, physical versions of mathematical puzzles, and tactile mathematical games. Polyhedra and tensegrity structures hung from the ceiling, a half-meter Soma Cube beckoned from a table, and sets of tiles to be arranged by satisfying various matching rules challenged visitors. Tens of thousands of Long Island schoolchildren visited for programs and activities over its two and a half decades of operation.

However, the Goudreau Museum's ultimate level of success suffered from a variety of obstacles. The principal founder passed away in 1985, just a few years after its opening. Its small size limited its offerings, and in turn, its geographical reach. Although its original group of founders remained loyal to the cause for many years, few new individuals were attracted to its governing board, and fundraising proved to be an ongoing struggle. In 2006, the Goudreau could no longer pay even its modest rent, and except for a few items retrieved by Goudreau's former colleagues, all of its collections and records were discarded.

Mission and Vision

Against the backdrop of these prior exhibitions, the Working Group came together. Most members had visited one of the copies of *Mathematica*. The principal founder of the effort convened the group in direct response to the news that the Goudreau Museum had closed, and two other initial members of the Working Group had been on the governing board of the Goudreau. (Over the course of opening the museum, leaders of the effort also visited other math museums, such as the Mathematikum in Giessen and Florence's Il Giardino Di Archimede, and met with the designers of other American math exhibitions, such as the Oregon Museum of Science and Industry's "Design Zone" and "Moneyville".) The overall sensibility of the group, emerging from this background, was that (American) society suffered from a dearth of opportunities to explore math through play and physical experiment, rather than by formal symbolic exercises and computation.

With the *Math Midway* having attracted thousands of visitors for its brief debut, the Working Group, and increasingly the new governing Board of Trustees, had to determine how to follow up that opening act. In fact, part of this decision had already been made in the leadup to the WSF. Conversations with festival staff and other potential participants made it clear that the name "Math Factory" gave a vague impression, at best, of what the organization would be and do. On the other hand, the Board felt that whatever its connotations might be, the word "museum" made it clear to all that, at least, this would be a place for the public to go and see interesting things. On this basis, the organization obtained, in late 2009, a charter from the NY State Department of Education as the Museum of Mathematics (MoMath) — the "National" would only be added by petition a few years later as the Board and programs of MoMath expanded their geographic reach.

So, if the goal was to create a place for people to come, it still remained to determine how that site would be organized, what would be in it, what activities would be possible there, and perhaps most importantly, what specifically it would try to accomplish for its visitors. The mission statement adopted earlier gives only the highest-level hints at these aspects.

Indeed, many involved in the project were personally motivated in large part by frustration with common societal views and attitudes about mathematics. These views often share a negative aspect and are probably familiar to many readers of this *handbook*; indeed, Chapter 6 of this book begins its second section with a litany of them. To the ones mentioned there, the Working Group might add: Math is only about numbers, consisting of rote exercises with only one predetermined set of steps to solve. If you don't happen to be a "math person", you'll never succeed and should just forget about math. The list, unfortunately, goes on and on from these highlights.

However, many of MoMath's advisors, especially those from a fundraising background, cautioned against trying to galvanize enthusiasm, participation, and support around negative propositions. They suggested that organizing MoMath's efforts and messages around the positive aspects of mathematics and constructive goals would more effectively advance its mission. Hence, MoMath shifted its orientation toward demonstrating and evoking new perceptions: Math is creative, with new ideas devised every day. Math connects to almost everything in the world around us. Math is open-ended. Math is universal and everyone can be a part of it, regardless of their background (or race or gender). Math includes an untold number of unsolved mysteries waiting to be explored. Math has an almost limitless array of different areas, so there's something for everyone. Beyond just these reversals of common misconceptions lie altogether new, positive viewpoints. Math is not only beautiful, but actively seeks beauty. Math is about patterns and reveals hidden structures. Math is created by people to better understand the world.

These positive formulations also provide much more powerful guidance in selecting exhibits and activities for the museum: If you want to show that mathematics is open-ended, choose activities that foster exploratory play rather than a preset script of actions. If you want to show that math produces new discoveries and has unsolved mysteries, find exhibits and programs that highlight recent results and relate to open questions. To reveal math's aesthetic side, make sure you have beautiful exhibits, where the beauty comes from the underlying math. Moreover, it's these positive aspects of math that attracted many of its professional participants. Focusing on them helps keep the museum interesting to mathematicians as well as the general public, providing a valuable source of ideas, collaboration, and goodwill for the enterprise.

Implementation

Numerous nuts-and-bolts issues remained between MoMath and its opening day, about which principles like "show how math seeks beauty" only seem to provide tangential guidance, if any. Probably foremost among these was the question of location: where would the museum initially open? Following the principle that few plays debut on Broadway, the Working Group initially pursued a location in Stony Brook, NY, which seemed favorable given the proximity to the highly regarded mathematics department at Stony Brook University. However, as the Board of Trustees grew, sentiment gradually arose that the potential visibility and attendance of the museum would be much greater in New York City. So why not attempt to get more social impact per unit of effort by locating there? This sentiment was clinched by a challenge grant from the Simons Foundation, which promised significant additional funding if MoMath could raise enough to open in New York City as well as meet a size threshold in the vicinity of 1,000 m² of exhibit area.

This new mandate forced the question of how MoMath would develop the operational experience and public following needed to open in a highly visible location while maintaining a professional level of quality. The two primary answers were the *Math Midway* traveling exhibition tour and the *Math Encounters* public presentation series.

After the WSF, the Working Group sought to realize the "enduring value" from the *Midway* that it had in mind when creating it, by identifying venues interested in hosting it. An immediate difficulty arose: most museums and science centers plan their exhibition schedules at least 2 years in advance. MoMath filled this gap with creative arrangements such as its first post-festival showing at the Urban Academy, a charter high school in Manhattan. Although the Academy could not pay a rental fee, it provided free space for the exhibition. The engagement allowed MoMath to gain experience with installing and striking the *Midway*, to develop lesson plans, and to establish a track record for making pitches to other venues. Sheer doggedness played a role as well. The chief operating officer (later CEO) of MoMath stopped at every science center she passed on a family vacation, eventually landing the exhibition's first paid rental, at the Da Vinci Science Center in Allentown, PA. Ultimately, the *Math Midway* appeared at twenty different public venues, culminating with a sale to the Singapore Science Centre in 2016. Each stop provided an opportunity to visit and learn from the operations of a different institution, meet with its leaders, and gain insight into the museum business.

However, at least after it left the Urban Academy, the *Midway* did not help with another important operational goal: developing a local following and constituency, who would generate interest and spur attendance once the museum opened. For that, MoMath created the *Math Encounters* public presentation series, also graciously funded by the Simons Foundation.

This program continues to date; once each month MoMath brings a gifted public communicator of mathematical ideas, often but not necessarily a professional mathematician or math-adjacent scientist, for a general-audience public talk [8].

In the spirit of a hands-on museum, most *Math Encounters* presentations include an audience-participatory component, to create active engagement with the topic and promote the perspective that anyone can do math. Although the primary criterion for presenter selection is an ability to reach and connect with a broad audience in an engaging way, programming for the series also seeks to highlight the incredible diversity of ways that mathematics touches upon our lives and our world: making sports teams better, gauging the shape of the universe, finding the best way to choose lottery tickets, detecting art forgeries, or devising new drugs, as a few examples. Thanks to the size of the market, New York's culture of public speaking events (exemplified by such institutions as the 92nd Street Y), and the scarcity of similar content, the series drew strong attendance from its inception, typically two to four hundred people in two sessions. Offering the series also provided another important source of operational experience, including such processes as operating a ticketing system, handling public visitation, and so on.

The new museum also required a collection of permanent exhibits. Fortunately, MoMath had more time for this second round of design, and so could adopt a more structured process. That process began by looking at a museum visit as a whole. What possible benefits, opportunities, or outcomes would MoMath like to afford its visitors as the overall result of their museum experience? To embody this question in the exhibition design, the four-person team at MoMath overseeing that process (consisting of a designer, an operations person, and two content specialists) chose to rephrase it as "what question will the overall permanent exhibition of the new museum answer?" Hopefully, a clear "guiding question" would ensure coherence among the individual exhibits and transmission of a unified message to the visitor.

It proved difficult to formulate an effective guiding question, despite a sustained effort. The exhibition team discarded some obviously reasonable and desirable questions, such as "What is mathematics?" or "What does it mean to do mathematics?" as having answers that are too comprehensive or murky. The question would do no good if it required more space than the institution could possibly afford in Manhattan even if fundraising exceeded expectations, or if it would not provide an organizing scaffold for exhibit selection. Other questions, like "Is mathematics a science?" were important, but seemed more appropriate for a math section of a larger institution, not a self-contained question for this institution. Still others, that seemed very attractive for MoMath's mission, such as "Why do people create mathematics?" appeared daunting to the exhibition team to translate into the types of exhibits they were familiar and comfortable with: manipulable physical objects embodying or illustrating a specific mathematical concept. In the end, the team settled on the somewhat unfocused "What are some of the most important concepts you encounter in the world of mathematics?" The idea was that the answers to this question, such as dimension, curvature, pattern, beauty, number, proof, probability, and so on, could serve as unifying themes for spatially-organized groups of exhibitions. Unfortunately, this choice did not afford a particularly crisp narrative informing a museum visit.

The actual exhibition design and selection process began by returning to the crowd sourced repository of exhibit concepts, which had grown considerably in the time since the *Math Midway*, and proactively seeking out other sources of ideas (including a visit to the legendary math communicator, Martin Gardner, two of whose concepts were realized by MoMath's opening day). In all, roughly four hundred concepts were identified to the extent of a title and a one-to-three sentence description. MoMath convened its Advisory Council, consisting of roughly two dozen math professionals, including pure and applied mathematicians, teachers, mathematical artists, statisticians, etc. The advisors spent an afternoon discussing the concepts and rating them. MoMath relied primarily on their "gut feelings", guided by the following priorities: first, exhibits should provide engaging, fun, and active experiences for the visitors; second, they should embody some aspect of mathematics in a substantive way; and third, they should be explicable and of interest to as broad a range of mathematical sophistication as possible (ideally ranging from someone with a fifth-grade math background all the way to a professional mathematician).

MoMath's designers drew up concept sketches for the top quartile of designs from the Advisory Council. The full exhibition team collaborated to extend and detail the description of the activity each exhibit would afford. This step filtered out a significant fraction of the candidates that nobody had a plausible idea of how to execute. One example: a sheet of flexible material that would automatically change color at every point to indicate the local curvature there. Such a sheet would clearly be fun to play with and illustrate an important mathematical concept, but has so far defied implementation. (We can imagine it will one day be possible with new materials like "smart fabrics.")

The exhibition team then hashed over each remaining exhibit idea, refining the concept further and deciding whether to move forward with it. Several of the *Math Midway* ideas had seemed good "on paper" but proved ineffective with, or unattractive to, actual visitors. However, MoMath lacked a facility or opportunity for a prototyping phase to "play-test" any but a few of the concepts with outsiders, to avoid such pitfalls. Cognizant of these difficulties, the team determined that only ideas that every one of the four members felt completely confident in would proceed, hoping to minimize the number of unsuccessful exhibits. However, in effect, this process gave each member of the team a veto on every exhibit, with a couple of unintended side effects: stifling some more unusual ideas that had just one or two strong proponents, and occasionally causing the debate on the merits of an exhibit to turn toward its negative, rather than positive, aspects. If pursuing a similar design process in the future, making room for each key participant to advance a limited number of designs solely on their individual approval might mitigate some of these side effects.

Note that the governing board of the museum also served as an important reality check in this process. Excited by the large array of promising exhibit ideas, and feeling a need to show MoMath's ability to generate novel content, at one point the exhibition team determined not to repeat any of the items from the *Math Midway*. The Board's response: "You think it's a good idea to exclude an iconic exhibit [the square-wheeled tricycle] that has consistently drawn hundreds of visitors and been written up in over a half-dozen languages? Please review that thinking." In the end, the tricycle, the laser-cross-section, and the brachistocrone exhibits from the *Midway* were reprised in the permanent exhibition.

Out of these deliberations came between 50 and 60 exhibits which were developed to the point of submission to fabricators for soliciting cost quotations. Budgetary and space considerations, as the designers began to lay out floor plans for their newly-rented space, winnowed these candidates to the slightly over 40 that have appeared on MoMath's primary exhibition floor to date. (See MoMath's website [9] for a complete list of current exhibits.) It should be noted that bringing the design team in-house allowed the museum to develop a very eclectic collection of exhibit styles, with references to art, architecture, and history, and a level of polish hopefully suitable for Manhattan's culturally sophisticated audiences.

One casualty of this final process was the original guiding question informing the design process. Some of the "key concepts" identified earlier had in the pragmatic winnowing fallen to one or two related exhibits. The physical realities of the space also allowed more of the approved concepts to fit when conceptual grouping was not enforced. Opinions on the exhibition team varied as to the relative importance of more exhibits versus establishing some higher-level organization, and ultimately the museum simply presented a collection of stand-alone activities. (Like the alphabetic positioning of the "fgh-jkl" keys on a QWERTY keyboard, an echo of the original organization remains in MoMath today: the grouping of *Wall of Fire*, *In Plane Sight*, and *3-D Doodle* in the northwest corner of level–1 stems from the "Dimension" response to the guiding question.)

It's also worth noting the enduring influence across five decades of the Eames *Mathematica* exhibition. Almost every one of the anchor exhibits of *Mathematica* has a counterpart in MoMath that seeks to refresh or extend its archetype. The train on the Möbius strip has become *Twisted Thruway*, which provides live first-person point-of-view video of the trip. The Probability Machine begat *Edge FX*, where visitors can alter the left-right probability at all of the pins and observe how it affects the resulting distribution. Visitors familiar with *Mathematica* will find numerous other parallels.

With the experience gained by the operational team and a considerable array of newly designed and fabricated exhibits, MoMath was ready for its December 2012 opening to the public. The focus here on the physical aspects of the museum shouldn't detract from the array of programs (beyond *Math Encounters*) which also bring the exhibits and mission of MoMath to life. Although space precludes going into detail, here's a brief selection: a series of shows straddling mathematics and art in the museum's temporary exhibition gallery; summer day camps for kids to explore mathematics; Friday night family math activities; annual scholarly conferences (alternating years between recreational mathematics and public math engagement); large-scale outdoor "math rallies" in celebration of Pi day and other special dates such as "Pythagoras day", December 5, 2013; and many others.

Operating Realities and Assessment

Of course, opening a museum is only the beginning of a story, as attested by the unfortunate recurrence of museums closing before they reach their second anniversaries. Clearly, day-to-day museum operations are hard. There's significant knowledge and experience in public communication of science to be gained in every one of the years that follow opening day. And to improve, an institution has to determine whether, and how well, it is accomplishing its mission and goals.

The blunt tools of measuring impact are straightforward. First, look at absolute attendance and attendance trends. If attendance is meeting projections, then you're at least having the opportunity to convey your messages. If attendance is increasing, it suggests that you're making some sort of positive impression on the people who come (even if it's difficult to tell whether or not it's the type of impression you wanted to make).

Second, in this internet era, it's easy to see what people are saying about you. Although the information gleaned in this way is unscientific and surely suffers from significant and unmeasurable selection and other biases, it can nevertheless provide useful feedback to guide the outreach mission. For example, on any site with reviews of MoMath, one can find visitor concerns that "many exhibits don't work". From an internal perspective, the dates of such comments can be correlated with maintenance records to find that whenever as few as two or three exhibits (out of over 40) are perceptibly non-operational, the museum becomes susceptible to such comments. Looking at this situation from the perspective of the visitor, we learn that non-operational exhibits distract from, and therefore pose a disproportionate obstacle to, the communication effectiveness of other exhibits and the museum as a whole. Clearly, a hands-on informal science institution must place a very high priority on exhibit operations in order to execute its mission effectively.

If it's so easy to glean that principle, you might wonder why it isn't the case on a daily basis that every exhibit is either working or inconspicuously removed or temporarily replaced. Isn't it simply a matter of devoting the necessary resources? Two major issues work against implementing a seemingly straightforward solution. First, it's impossible to make exhibits visitor-proof. In their process of exploration, attendees push and prod exhibits in ways that designers can't possibly anticipate, applying stresses and forces that have an uncanny knack of finding points of failure. Of course, as science communicators, we *want* visitors to explore in unexpected ways. So breakage remains inevitable. The most effective countermeasure consists of designs that incorporate easy-to-repair elements that will break in lieu of more intricate, sensitive components. However, without an extensive prototyping phase — "building one to throw away", so to speak — it's very difficult to execute such designs *ab initio*. As a result, MoMath continues its efforts to redesign exhibits to reduce downtime to this day.

The second issue is simply the importance of staff resources in museum operations, especially technical staff given the complexity of MoMath's exhibits. Primary floor staff of course play a critical role in communicating the meaning and ideas behind activities, but they also serve as the front line in ensuring that all visitor-facing items work smoothly. Many problems require additional troubleshooting and repair, which needs specialized expertise. On the other hand, funds for compensation at a nonprofit can be tight, the necessary work can become repetitive, and other industries like technology and finance may offer higher compensation for technically-skilled museum employees. These factors can contribute to staff turnover, compounding operational difficulties, and potentially posing a risk to institutional memory. Thus, creating other positive aspects of remaining in the organization, such as empowering staff to solve problems creatively and implement new ideas to enhance visitor experiences, becomes critical to success in communicating science.

Visitor comments also reveal a less frequent, but persistent, sentiment that interacting with the exhibits is fun, but that it's unclear what the experiences add up to. Ironically, MoMath's deliberate orientation toward highlighting the breadth of mathematical inquiry, stringently limiting the number of exhibits that make any mention of numbers in particular, may exacerbate this issue by depriving visitors of places to anchor their pre-existing, likely number-heavy, conceptions of math. Still, comments like "I hate to tell you, your activities are fun, but there's no mathematics in this museum" help crystallize a recurring phenomenon in mathematics exhibitions, discernible even just among the ones covered in this article. *Mathematica*, the Goudreau Museum, the *Math Midway*, and MoMath's initial permanent exhibition all ended up as eclectic groupings of disparate subtopics and types of activities. Perhaps stemming from an ongoing struggle to generate a critical mass of ways to translate the abstract concepts of mathematics into engaging physical activities, this character of the exhibitions can well convey the intellectual variety found in math, but may make it difficult for visitors to perceive or synthesize broader themes or aspects of the subject. This type of feedback also validates the original design aspiration of providing an overarching narrative for a museum visit, and suggests that future projects in mathematics communication are well-advised to pay attention to the coherence and cumulative message of their individual parts and topics.

Ultimately, honing communication effectiveness requires more substantive, rigorous assessment. Recognizing this, MoMath has commissioned at least one professional evaluation by an independent external firm. Such studies can help quantify the true prevalence of concerns that show up in informal feedback, as well help to understand more nuanced issues such as the effect of a museum visit on attendees' attitudes toward mathematics. Although the results of the specific evaluation mentioned are accessible only to the museum's administration and so can't be analyzed here in detail, they will surely help MoMath gauge both its strengths and opportunities for improvement.

MoMath also attempted to build assessment processes into its core design. For example, one staple of museum exhibition evaluation consists of visitor tracking, to determine such statistics as dwell time per exhibit, percentage of exhibits visited, and so on [4]. For many years following its opening, MoMath used RFID visitor badges (to indicate admission had been paid) whose location could be automatically determined by sensors throughout the museum. Another common technique administers brief exit questionnaires to visitors, and MoMath installed a kiosk to give attendees the opportunity to provide such feedback. In theory, it would even be possible to correlate times and trajectories of visitors with responses to the questionnaires. A considerable amount of data was collected from both of these sources, but to date there has not been any fruitful analysis of such data.

In a similar vein, MoMath intended to incorporate some of its core messages into the fabric of its permanent exhibition by technological means. All exhibit guidance texts and written interpretive information are presented on electronic displays rather than printed panels. This way, visitors could effortlessly access explanations at multiple levels and in whatever language they are most comfortable reading, helping to convey the universality and inclusivity of mathematics. In practice, while some visitors may take advantage of selecting interpretations available on most exhibits at three different levels of mathematical sophistication, marshaling the effort to create the necessary texts in even a second language has proven beyond the reach of limited staff.

In part because of historical difficulties with inclusion in the professional mathematics community, MoMath has also faced the challenge of conveying that message via the public face of mathematics it presents. Examples of such difficulties include the following: of the first 113 speakers in its flagship *Math Encounters* series, 78% have been male and 81% have been white; and the time devoted to a museum educator-led session during a donor-subsidized school visit (which typically serves a more diverse population) has been only two-thirds of the time given to schools that purchase such sessions on their own.

Recently MoMath has made significant efforts to address such challenges. New programs focus on encouraging young women and minorities to engage with math. For example, *The Limit Does Not Exist* is a series aimed at girls 10–15 years of age in which panels of accomplished female mathematicians discuss their personal career journeys in a relaxed setting, and *Bending the Arc* brings Black youth together with leading Black STEM professionals. A recent project added the names and faces of women to many of the electronic exhibit displays, with 33 women now mentioned. Gender representation appears to be improving in *Math Encounters* as well: nine of the last twenty presenters have been female. That said, these efforts are only a start; much work remains to be done in this regard both at MoMath and in the mathematical community at large.

The operational difficulties and realities sketched in this section should, however, not obscure the signs of success on MoMath's part. Its cumulative attendance has topped one million visitors. Average ratings on review sites remain steadily high, and comments make it clear that for a large segment of visitors, MoMath accomplishes at least its affective goal of enabling them to enjoy activities associated with math. Moreover, it's created a place where it's safe to express and take pride in positive views and sentiments toward mathematics, which sadly isn't always the case in other social settings. It's won numerous awards and accolades, and significantly increased the number and variety of hands-on mathematics activities available to the public. Along the way, it's produced hundreds of hours of math videos, attracted hundreds of thousands to public events, and created numerous resources for teachers, students, and other budding lovers of mathematics. MoMath can also serve as an inspiration and support for other math outreach organizations, strengthening their efforts through sharing its experiences. Hopefully many others will reap the sort of reward that the MoMath creators were fortunate to experience early on: the joy of seeing a young girl tugging on her mother's arm, exclaiming "Ooh, Mommy, there's *math* in there!", eyes aglow with wonder.

Acknowledgments

The author wishes to thank Sylvie Benzoni and Marion Liewig of the Institut Henri Poincaré, Cindy Lawrence and Tim Nissen of the National Museum of Mathematics, and the volume editors for their helpful comments and suggestions in preparing this chapter.

References

[1] Dawson, E. (2014). "'Not Designed for Us': How Science Museums and Science Centers Socially Exclude Low-Income, Minority Ethnic Groups". *Science Education*, 98(6), 981–1008.

[2] Wagon, S., and Hall, L. (1992). "Roads and Wheels". *Mathematics Magazine*, 65. Doi: 10.2307/2691240.

[3] Whitney, G. T., Hart, G., Lawrence, C., Nissen, T., and Fan, C. K. (2016). "Square Wheel Tricycle". US Patent No. 9,472,120 B1, October 18, 2016.

[4] Constantine, E., Frankfeldt, H., Low, and Schoenfeld. (2004), "Summative Evaluation of Mathematica: A World of Numbers and Beyond".

[5] Tscherny, G. (1962). *Mathematica: A World of Numbers and Beyond*. Armonk, NY: IBM.

[6] "Galton Board", https://galtonboard.com. [Accessed on December 3, 2020].

[7] Mirenberg, K. J., "Introduction to Gravity-Well Models of Celestial Objects". https://www.spiralwishingwells.com/guide/Gravity_Wells_Mirenberg.pdf. [Accesed on December 5, 2020].

[8] *Math Encounters*. video archive. https://youtube.momath.org.

[9] "MoMath Exhibit Guide", https://momath.org/explore/exhibits/. [Accessed on December 13, 2020].

© 2023 World Scientific Publishing Co. Pte. Ltd.
https://doi.org/10.1142/9789811253072_0007

Chapter 6

Bridging the Gap Between Math and Society

Sylvie Benzoni* and Marion Liewig[†]

Institut Henri Poincaré, Paris, France
*sylvie.benzoni@ihp.fr
†marion.liewig@gmail.com

Context

The *Maison Poincaré* is an emerging center for the public exploration of math, created by the Institut Henri Poincaré (IHP) in response to a crisis in French society's interest in and comfort with mathematics.

The IHP itself is a research and training center founded in 1928 [6]. It will be hosted by 2022 in a refurbished building.

The district where the IHP is located, and thus where the future Maison Poincaré will be, is mostly frequented by university students, intellectuals, and researchers. Few tourists, families, or schools actually visit that district. One of the main goals of the Maison Poincaré is to attract and open doors to a large variety of people, ranging from high school students and teachers to occupational users of mathematics, art amateurs, and laymen. Of course we hope that it will also attract people already keen on mathematics, but this is not the main purpose.

The project has arisen in a rather paradoxical context. France is praised for the excellence of its mathematical elite, but on average pupils and students perform poorly in international assessment surveys.[a] Math curricula have been revised over the years, but the situation has worsened.

It turns out that the latest reform was based on a report [1] coauthored by the IHP former director and initiator of the Maison Poincaré project, Cédric Villani.

---

[a] PISA https://www.oecd.org/pisa/publications/pisa-2018-results.htm; TIMSS https://timssandpirls.bc.edu/timss2019/.

Despite its good intentions, that reform has had the effect of drastically reducing the number of high school students learning mathematics.[b] Math options in high school are perceived as too difficult, and students refrain from choosing them in dramatic proportions. Unfortunately, this effect is even worse on girls, who are traditionally more reluctant than boys to engage in math in Western societies and in France in particular.[c]

Nevertheless, secondary schools desperately need math teachers. This occupation is not attractive enough. The hiring of math teachers has been difficult for many years, with too few candidates, and candidates who do not meet basic requirements.[d]

Last but not least, a fact that is apparently not specific to France is that many people — including policy makers — proudly claim they are very bad at math, in a world where math has never been so ubiquitous. This has been emphasized by the COVID-19 crisis, which has shown severe deficiencies in the general level of understanding of basic mathematics.

This dire context gives us a strong motivation for building the Maison Poincaré. We undertake the challenge of contributing to reverse the situation from several perspectives.

There are scores of specialists in the world working on ways of bridging the gap between math and society. We may think for instance of the two concurrent series of international conferences, the Mathematics Education and Society Conference,[e] and the International Congress on Mathematical Education (Gueudet *et al*).[f]

---

[b] This has been reported to policy makers by various associations and learned societies, e.g., the SMF https://smf.emath.fr/actualites-smf/0320-lettre-ouverte-cpu-reforme-voie-generale-lycee-enseignement-mathematiques, and in the general media, see e.g., https://www.nouvelobs.com/education/20201127.OBS36684/info-obs-comment-la-reforme-du-lycee-a-eu-la-peau-des-maths.html.

[c] The proportion of girls choosing math in their last high school year has been divided by 2 after the latest reform. More precise figures are analyzed on https://femmes-et-maths.fr/2022/01/25/reforme-du-lycee-25-ans-de-recul-sur-les-inegalites-filles-garcons-en-maths/.

[d] For the last decade, the national competition for hiring secondary school math teachers, CAPES, has been recruiting fewer people than the number of open positions, see https://capes-math.org/data/uploads/Reunion_formateurs_2020.pdf.

[e] https://www.mescommunity.info/.

[f] https://www.mathunion.org/icmi/conferences/icme-international-congress-mathematical-education.

At IHP we are happy enough to have acquired the means to take action in mathematical science communication, if not directly in mathematics education. We aim at implementing various ways of developing *mathematical acculturation*, in the most positive sense of this term.

The Maison Poincaré will include a permanent exhibition, to which the next part is devoted, and various outreach activities in connection with it, to be set up in our premises or outside the walls in collaboration with our partners. Our experience of outreach activities so far and some plans in this regard are described in section "Outreach Activities" (page 106). Finally, shorter parts are devoted to discussion of our assets (section "Assets", page 110) and our challenges (section "Challenges", page 112).

The Permanent Exhibition

Ambition and target

Numerous myths surround mathematics. Even when debunked, see, for example, [2], math myths persist in society and in the students' minds. Let's quote a few of them, which we think contribute a lot to the bad reputation of mathematics. Math require arid logic, not intuition.[g] Math is not creative. There is nothing new to discover in math, its knowledge body has been stuck since the antiquity (Pythagoras)/the 8th century (Al-Khawarizmi)/the 17th century (Fermat/Newton).[h] Math is useless, it has nothing to do with daily life. Not to mention the all-too-common misconception that men are better at math than women.[i]

We aim at deconstructing the bad reputation of mathematics in various ways. We plan to do so in particular by opening up the public to hidden and/or surprising mathematics on a variety of topics than can pique their interest, for instance, starting with objects from daily life (e.g., soccer ball) or situations (e.g., crowd evacuation).

---

[g] Debunking this myth is one of the main goals of the recent book by David Bessis — *Mathematica: Une aventure au cœur de nous-mêmes*, Seuil 2022.

[h] The vast majority of people would choose the first option, believing that math stopped with Pythagoras.

[i] This belief is rooted in even worse ones that were shared by the most prominent scientists until very recently, see e.g., this quote by the physicist Richard Feynman in 1966: "I didn't realize the female mind was capable of understanding analytic geometry." (http://www.feynman.com/science/what-is-science/).

Our ambition is to convince visitors that mathematics and their applications form a huge, lively, and crucial field of science and technology. This is important to have more young people engaged in this field and related ones. This can help to comfort mathematics teachers in their job [3].[j]

On a more general ground, we want to demonstrate to visitors that mathematical thinking is of great help in understanding the world, and also for making decisions in daily life. Anyone can gain from it. Anyone can belong, including of course girls and women, and people of diverse backgrounds.

The main point of originality underlying this project, compared to other science or math museums, is that it is supported and hosted by a research institute.[k] It is important to our supporting institutions (CNRS and Sorbonne Université), other public funders (the city of Paris, the Ile de France region, the French state), and our private partners through the IHP endowment fund, that we convey research topics, and that we bridge the gap between researchers and the public.

Visitors will have the opportunity to meet and exchange with researchers hosted at IHP. Besides the classical format of public lectures, researchers will be encouraged to interact with the public through a renewed tradition of tea parties in Perrin's former tearoom [4]. The café by the entrance, the multiple terraces of the building, and the garden equipped with a blackboard and a full-scale version of the Maison Poincaré symbol — a mathematical sculpture called Rulpidon — will also be pleasant places for researchers and the public to meet and chat. This will enhance the role of the institute in the society, and participate in our endeavour to demystify mathematics.

---

[j] To quote Bishop (1998), "in my experience there are few teachers of Mathematics [...] who consider that they are themselves Mathematicians [...], or even that the Mathematics which they teach has any particular values. [...] They could be acculturating their students into believing that they are learning Mathematics for its benefits as applicable knowledge, or to have pleasure in its finest or most intriguing discoveries or inventions, 'for its own sake', or even to train their minds."

[k] We are not aware of any other example, save for the small Museum für Mineralien und Mathematik Oberwolfach linked to but not exactly in the premises of the Mathematisches Forschungsinstituts Oberwolfach (https://www.mima.museum/?lang=en), even though many institutes do offer outreach activities (e.g., the Perimeter Institute https://www.perimeterinstitute.ca/outreach).

Our purpose is to reach the widest possible audience from the age of 14, from the most reluctant people to the most enthusiastic ones, making them discover what research in mathematics and applications is like, and allowing them to encounter the men and women working in those fields. We have been working with associations and the three regional academy directorates (Créteil-Paris-Versailles) to get in touch with students from diverse backgrounds and their teachers. We have also been working with art schools, museums, and a movie theatre (see section "Partnerships", page 112, for a few more details). We communicate on a regular basis on social networks (Facebook, Instagram, Twitter, YouTube), and send newsletters to our several thousand entries database, drawing attention to our outreach activities.

In doing so, we have the ambition of contributing to raise public awareness of the importance of mathematics for contemporary society and to encourage more young people to engage in mathematics and their applications in technology and society. In order to achieve this goal, we need to convince the public that mathematics is a vivid field with lots of occupational opportunities, and make this visible in the media and social networks.

This will be done in particular by drawing attention to men and women with diverse trajectories that have been strongly influenced by mathematics, and by pointing out unexpected, close links between academic research, applied research, and innovation. If not as playful as other math/science museums, our permanent exhibition is intended to be inclusive and induce astonishment, to arouse curiosity, to raise questions, and create wonder.

Our main target audience is the population of high school students,[l] teachers,[m] and those people in the general public who have an interest for science, even more so if their interest comes from indirect paths — for example, through art or humanities. We have acquired some experience in this respect, as we explain below (section "Assets", page 110).

---

[l] And also advanced middle school students.

[m] We would in particular like to attract teachers from other disciplines or primary school teachers who were not originally trained at science.

We shall pay special attention to targeting less privileged zones and to attracting schools from priority education zones, as we have already done in various outreach initiatives. For instance, we have organized visits and public lectures with the *Institut Télémaque* (mentioned in section "Partnerships", page 112). Furthermore, the IHP director Sylvie Benzoni is engaged in a training program for math teachers of the whole Créteil-Paris-Versailles academic region.

She has also been building special links with a middle school in Saint-Denis (northern suburb of Paris), which has chosen her as a role model among other female scientists. The students of that school will certainly be the first to test some of the Maison Poincaré exhibits hands-on. We are going to work with the association *Science Ouverte* (also mentioned in section "Partnerships", page 112) to open this possibility to more schools from that area.

We are also convinced that we have a role to play in conveying scientific culture more broadly. In this we rely in particular on journalists, science journalists but more importantly those from the general, national media (e.g., *Le Monde*) or cultural ones (e.g., *Beaux Arts Magazine*). We already interact with a number of them on a regular basis, thus providing help to enlarge their scientific network and contributing to spreading the word about scientific stakes. We plan to reinforce initiatives targetting the media in the Maison Poincaré.

Finally, we aim to involve the international visitors participating in IHP research programs more with outreach efforts at the Maison Poincaré. Some of them are already willing to contribute to raising public awareness, while others need to be encouraged, as we have experienced in the last couple of years when asking for public lectures. Encouragement might come, for example, from specific additional financial support to stay at IHP and engage in a mathematical science communication project.

Design method

At first the permanent exhibition project was being discussed between the IHP directors — Cédric Villani, Jean-Philippe Uzan and then Sylvie Benzoni, the project manager Marion Liewig, and the scenographer, Rémi Dumas, working with the architect in charge of the building's refurbishment.

The need for a curator was soon identified if we were to make any progress in designing the contents of the museum's program. This is how Céline Nadal, a museographer originally trained at advanced physics came into play. She started to work for the project in 2019 and she is to continue until its opening in 2022. Her role is most important to lead the workgroups and adapt for the public the content imagined and/or produced by the researchers involved in the project.

Before and while working on the museum's program, we visited several science museums at various places in the world to get some inspiration, references, and points of comparison. Visits were made abroad at the *MoMath* in New York, the *London Science Museum*, the *Manchester Science Museum*, and the *Mathematikum* in Giessen. In France we are in close contact with the *Palais de la découverte*, the *Musée Curie* and the *Espace des sciences Pierre-Gilles de Gennes* (ESPGG) in Paris, the *Maison des mathématiques et de l'informatique* (MMI) in Lyon, and *Maison de Fermat* near Toulouse. This has been helpful for us to better define our positioning and to think about our future exhibits. Some non-scientific museums such as the *Deutsche Kinemathek* in Berlin or the *Hélène and Édouard Leclerc pour la Culture* in Landerneau have also been great sources of inspiration.

Three major principles have been underlying the design of our permanent exhibition:

1. Displaying topical mathematics embodied by contemporary men and women involved in mathematics or connected fields.

2. Paying attention to gender equality, diversity, and accessibility.

3. Managing the project in a cooperative and collaborative manner.

In the spring of 2018 we already had a large committee called *Comité de culture mathématique* (CCM) at IHP. We enlarged it with volunteers who got interested in the project: researchers, teachers, science communicators, partner company representatives, etc. This enabled us to gather a wide spectrum of expertise and a critical taskforce as well. We have continued to involve more people when we felt the need for it.

Smaller workgroups have been set up for each space, sometimes specifically for a given exhibit. At least two active researchers take part in each

Table 1:   People figures.

1	Museographer
1	Scenographer
1	Graphic designer
56	Researchers
10	Teachers
8	Science communicators
3	Deeptech company representatives
6	Partner company representatives
5,000	Expected visitors per year — students not included
600	Expected high school classes per year
6	Staff

of those collaborative pieces of work. A few figures are collected in Table 1. We have been using this method for years to design smaller exhibitions and other outreach events. The institute is committed to maintaining this style of cooperation at least through the opening of the Maison Poincaré.

Planned visitor experience

The permanent exhibition has been devised to offer a unique experience for visitors. It will occupy the whole ground floor with no predefined visit route. Instead visitors will discover its various spaces characterized by an activity verb. In mere alphabet order they read: becoming; breathing; connecting; inventing; modeling; sharing; visualizing.

These action verbs are the skeleton of the museography program, intended to reveal various facets of what it means to practice mathematics. They have been chosen consistently with the layout of the premises and their historical significance. They will be highlighted on suspended, transparent signs in French as summarized in Table 2. All other signs and panels will be translated into English so that foreign visitors can also enjoy visiting.

The exhibition will involve a great variety of museographic devices and appealing formats, ranging from hands-on displays, text panels, storytelling, portraits, and videos to 3D printed objects, interactive screens, augmented reality, and artwork. It will also include part of the

Table 2: Spaces and verbs correspondence.

Foyer	CONNECTER	Connecting
Gallery	MODÉLISER	Modeling
Lecture hall	INVENTER	Inventing
Tearoom	PARTAGER	Sharing
Perrin's office	DEVENIR	Becoming
Alice's room	VISUALISER	Visualizing
Garden	RESPIRER	Breathing

Figure 1: Artist view of foyer, with map of mathematics on the right wall. © Atelier Novembre and Du&Ma.

institute's heritage collections, mathematical models [5], and calculating machines.

We cannot describe every feature of the exhibition here. We have chosen to focus on a few of them that are quite representative of the spirit of the museography program.

Focus on low-tech exhibits

Map of mathematics — Connecting different fields

One of our low-tech exhibits aims at showing that mathematics is very diverse, with many fields that are interconnected to each other as seen in Figure 1. It consists of a map surrounded by a certain number of more or less common objects that can be linked to mathematics. The map is a little bit like a metro map, with metro lines serving as metaphors for wide fields such as analysis or geometry and metro stations representing more specialized domains such as calculus of variations or differential geometry.

The public will have text panels to read that explain links of the reproduced objects (e.g., credit card, horse saddle, roulette, smartphone, etc.) with those domains. For those visiting on a guided tour (see section "Individual vs. Guided Tours", page 105, for more details on tours) our science communicators will have the possibility to connect those objects physically to one or more "metro lines" thanks to elastic strings, and tell them more about those connections.

The hands-on exhibit available in the middle of this space will have to do with several of the highlighted domains.

Coupled pendulums — Illustrating solitons

Another low-tech but nevertheless sophisticated exhibit is intended to display the concept of a nonlinear wave. It will be made of a series of coupled pendulums on which the public will be able to launch various kinds of waves. This is a mechanical model for observing the solitary waves that are otherwise ubiquitous in several domains of physics. We think that this is a great topic for our science communicators to talk about many applications of mathematics.

This "soliton" exhibit will be part of the gallery on the theme of modeling, which will offer many other devices for the public to explore.

Data separation — The basics of artificial intelligence

Also in the modeling gallery will be an exhibit representing what is known as the perceptron, a basic ingredient in artificial intelligence algorithms. The device will illustrate how some data can be classified in a binary manner and separated by a simple surface. As can be seen in Figure 2, the classified data will be physicalized by small, colored balls — here blue or red — in 3D. The public will be able to walk around the exhibit and see those blue and red balls either apparently mixed altogether or, if they choose the good observation point, well separated by a plane.

Figure 2: Artist view of the gallery. © Atelier Novembre and Du&Ma.

Some explanation will be given on the text panel. This exhibit will be a good introduction to the Holo-Math (see section "Holo-Math — Experiencing Brownian Motion", page 104) episode on artificial intelligence, which is to be developed after the one on Brownian motion.

Researchers' experience narrative

In the space devoted to the theme of sharing, one of the exhibits will consist of a narrative that the public can listen to under a sound shower while leafing through a book. This is not completely low-tech since an RFID tag will turn the audio on and off.

Researchers will have told a writer how they live their academic life and how they come up with new ideas and make discoveries possibly in unexpected situations. The overall story will have been written and illustrated by professionals, and the text read and recorded by an actor.

Videos on various forms of engagement

Also, not completely low-tech will be videos presented in Perrin's former office, on the theme of becoming. They will be short documentary films on people who have been recognized for their engagement of mathematics in society. They will be screened on either side of the chimney. Everywhere in the museum, there will be a balance between men and women.

Focus on high-tech exhibits

Interactive map — Math around the world at different ages

The chalkboard of the tearoom can go up and be hidden in the casing above it. It thus releases a free wall in the rear. On this wall will be projected a map on which visitors will be invited to navigate through a remote panel. This map will aim at showing documents and other visual pieces of mathematics around the world. A cursor will also allow them to navigate in time.

This interactive map is a modestly technical exhibit that will tell and showcase how mathematics appeared, and how it has been circulating and developing up to the most recent times. The specific contents will be worked out so as to be diverse, surprising, and illuminating. Several of them will echo the other exhibits available in the room, in particular the device showing and whispering formulas, and the showcase dedicated to the connections between art and mathematics.

Image processing — Beyond the Fourier transform

In the area dedicated to modeling, an electronic exhibit will enable visitors to apprehend what image processing means. They will have the possibility of compressing their own digital image, with a slider allowing them to choose the compression rate, according to two main processes, namely the Fourier transform and the wavelet transform. They will be invited to observe the advantages of the latter over the former on two separate screens. Some complementary explanation will be given on text panels.

Holo-Math — Experiencing Brownian motion

The Holo-Math project is undoubtedly the most innovative one to be presented in the permanent exhibition. Following an original idea of Villani, it consists in using mixed reality to enable the public to use their own bodies to experiment with abstract concepts. Up to our knowledge, this is a completely new way of addressing mathematical science communication.

The main features of the Holo-Math experience will be the immersion in a 3D-animated virtual universe (Figure 3) without losing visual connection

Figure 3: Artist view of the Holo-Math episode on Brownian motion. © Eddy Richard (One More).

with the physical environment, and the possibility of interacting both with this universe and with the other people around. This is what mixed reality headsets make possible. Another important feature is the crucial role of the science communicator who leads the experience.

The proof of concept was made in 2017, and a first 25-minutes episode is about to be achieved. It deals with Brownian motion, which is both a physical phenomenon[n] and a profound mathematical concept involved in the modern theory of probability and various applications.

This has demanded a tremendous amount of creative development. Given its huge cost, we are now seeking new resources to develop an episode on artificial intelligence.

Individual vs. guided tours

We plan to offer four ways of visiting the permanent exhibition. Individuals will be allowed to enter and visit on their own. Otherwise, there will be focused guided tours, fully guided tours, and private tours.

---

[n] A phenomenon that was used by Perrin to demonstrate the very existence of atoms, which earned him the Nobel Prize.

Focused guided tours will consist of a 20-minutes exchange with a science communicator. They will enable, for example, the visitors to go deeper in a specific scientific question or challenge, or discover the life of a famous mathematician.

Fully guided tours will be intended for groups, mainly school classes or groups of teachers preparing a class visit. Private tours will be organized on demand, as has been repeatedly requested by our financial partners.

In order to optimize attendance and facilitate financial management, guided tours will be charged prior to visits.

Supplementary material

We have not yet defined the range of supplementary material but this will certainly include booklets made available from the shop, and detailed content on a dedicated website. We are also working with existing websites like Images des mathématiques° to offer various resources online.

Outreach Activities

When the Maison Poincaré opens we will be ready to offer a diverse program of outreach events and activities to complement the permanent exhibition. Indeed, we have acquired a wide experience in this matter since 2016. We are going to keep up the momentum, and continue to experiment with various forms of science communication inside and outside the walls. We pay a lot of attention to public feedback in order to improve our offerings regarding mathematics popularization.

Researchers

We organize public lectures by researchers who are staying at IHP, intended for schoolchildren and students but also for the general public. One difficulty is that a majority of visiting researchers are foreigners who can deliver lectures in English but not in French, whereas the targeted audience is not comfortable with attending a conference in English. This situation is improving given the number of incentives to promote English learning in the population.

° http://images.math.cnrs.fr/.

During the 2020 COVID-19 health crisis, we experimented for the first time with the organization of a webinar. We had invited a panel of researchers on-site, we had a remote audience who could listen to them live through our brand new video broadcasting and recording system, and we enabled a conversation between researchers and the public through a chat managed by our animators. The feedback has been very positive and we are going to follow it up with more content in this style.

We also have a partnership with Images des mathématiques, a website dedicated to the popularization of mathematics by researchers. We actually contribute with interviews of our visiting researchers, who thus introduce the readership to current research topics. In 2021 for instance, we talked about evolution, phylogeny and ecology, gravitational waves, topology, and cryptography.

Exhibitions

Over the years the IHP team, together with their scientific network, have designed a number of temporary exhibitions, most of them being available for loan to other institutions. These include exhibitions with a historical

Figure 4:   Exhibition on Jean Perrin, IHP library in September 2020. © Camille Cier.

flavour in honour of past mathematicians[p] on the occasion of anniversaries (Figure 4), which we designed more for an informed public.

For the last couple of years, we have developed partnerships with MMI and Maison de Fermat to develop joint exhibitions that are more aimed at younger generations, in order to encourage them engage in STEM. For instance, we designed an exhibition on surfaces that was shown for one school year at the Musée des Arts et Métiers and then at MMI. An exhibition on randomness was designed and shown at MMI. Similarly, there is an exhibition on artificial intelligence (AI) prepared at Maison de Fermat and shown at MMI. Both will be shown later in the Maison Poincaré.

Films

In the same spirit, IHP has produced several documentary films that meet cinema standards, in complement with an exhibition, save for the last, most original film. The first ones are dedicated to a scientist (Einstein, Lagrange, Shannon) and some of his achievements (general relativity, rational mechanics, information theory). The last two films are more art-oriented. One is about a math enthusiast who carves surfaces on copper, Patrice Jeener. The other, *Man Ray and the Shakespearean Equations* plays with various views of the mathematical models pertaining to the institute that inspired the surrealist artist Man Ray. Two new films are in preparation, one in connection with a thematic research program and one about some societal issues linked with the use and misuse of mathematics.

Like for the exhibitions, we are happy to circulate documentary films as much as possible. There are DVDs that can be borrowed from our library, and screenings in remote institutions can be organized on demand, even abroad since the films are subtitled at least in English and can easily be subtitled in any other language.

Still in connection with the world of cinema, the institute has run a film club since 2014 at a nearby, legendary movie theatre nowadays called *Le Grand Action*. Each session consists of the screening of a movie linked

---

[p] Maurice Audin, Gaston Darboux, Albert Einstein, Joseph Fourier, Évariste Galois, Sophie Germain, Joseph-Louis Lagrange, Jean Perrin, Henri Poincaré, Claude Shannon, Alan Turing.

near or far to science followed by a recorded exchange[q] between the viewers and our invitees, who are generally researchers or film professionals.

Jams

In our desire to reach young and non-scientist people, we have also experimented with innovative formats. The most innovative one was set up in 2019 together with the learned societies hosted in the building. After its big success, another one should have taken place in 2020 if the pandemic had no disrupted in-person activities. We called it a "Jam".[r] The idea is to bring together researchers with laymen. They would gather for 2 days at the institute to imagine and produce new forms of creative expression on a given scientific theme.

The 2019 theme was chaos. Our Jam du chaos (Figure 5) brought together more than 50 participants from very different backgrounds, such as plastic artists, musicians, video artists, graphic designers, students,

Figure 5:   Jam du chaos, Hermite lecture hall at IHP in October 2019. © Camille Cier.

---

[q] Subsequently made available on the institute's YouTube channel.

[r] In English in the text.

developers, scientists, and science communicators. In 2020, the theme was going to be health, in partnership with INSERM.[s] Due to the health crisis, we had to postpone the Jam till 2021, and we organized a blended event instead, with a panel on-site that was filmed and live broadcast and a chat online.

Podcasts

Furthermore, we launched a podcast series in 2019 with a professional radio journalist, hoping to depart from traditional radio broadcasts even as we took inspiration from a few of them. We wanted to highlight the temperament and ideas of those who do contemporary mathematics irrespective of their being known by the public or having published popular books. All the more so if they were not and had not.

This monthly podcast series is called *L'oreille mathématique*,[t] which does not translate well to English. Each episode is about 30 minutes long and features a conversation between the journalist, an invited scientist, and a science communicator. Having had very good feedback so far, we have continued this project, still paying attention to the representation of women and to the diversity of profiles when choosing our guests.

Workshops

Finally, a project that had been postponed due to the COVID-19 crisis took place in 2021. It was a workshop at the crossroads of academic research and tech companies on the theme of artificial intelligence. This is a format that we plan to renew on different topics.

Assets

One of our main assets in building the Maison Poincaré is the experience described in section "Outreach Activities" (page 106). Fortunately, we can claim complementary assets that are as important. These have to do with the very identity of the institute, with leadership and management features, secured financing, and strong partnerships.

---

[s] The national institute for health and medical research.

[t] http://maison-poincare.fr/podcasts/.

Prestige

The Maison Poincaré is not to be created from scratch. It is being built within the IHP, a research institute that is already internationally recognized. This configuration facilitates the work for its visibility. Even though the institute's visibility has to be developed further toward the general public, it is really an asset for enhancing the dialogue between scientists and society. Scientists are present within our walls. It is up to us to go and convince them to interact with visitors.

Team

Another asset not to be overlooked is the steering and management of the project. For several years now, the institute's strategy has integrated this mission of disseminating science to society. As mathematicians and physicists, the past and current directors and deputy directors have in common the knowledge and a real experience of scientific communication.

The operational team, the IHP staff, is wholly committed to the project, which clearly ensures desirable continuity and stability. Among them, six people are thoroughly involved in science communication missions. Remarkably enough, the IHP library is playing an important role in in outreach activities, by contributing to and hosting, if not steering, exhibitions, by restoring and displaying the mathematical models, and by developing a documentary collection intended for a wide readership on science and its applications.

In addition, there are numerous skills at stake in the project team, namely, architecture, engineering, museography, scenography, audio-visual competence, digital skills, iconography and documentary research, project management, marketing, communication, financial management, partnerships, and networking. We can rely on them to respond to basically any emerging problem.

Support

We can also rely on, and are grateful to our funders for their continued support every year. Sorbonne Université and the CNRS follow the progress of the Maison Poincaré very closely. They are both proactive, and they have listened remarkably to our development needs. Our other public partners,[u] the IHP endowment fund, patrons and partner companies, and foundations are also powerful allies. They provide us with valuable advice on financing, networking, and expertise with regards to human resources, business models, marketing, strategy, and media.

For each of our outreach projects like documentary films, exhibitions, augmented reality developments, book publishing, we have been assisted by service providers and professionals who excel in their fields.

Partnerships

In recent years, we have developed links and fruitful partnerships with inspiring scientific and cultural organizations. This includes neighbouring ones like the *Musée Curie*, the *Panthéon* and *ESPGG*, other Paris museums like the *Musée des Arts et Métiers*, *Palais de la découverte*, *Musée du 11 Conti* (Monnaie de Paris), *Musée de l'Homme*, *Musée national des arts asiatiques - Guimet*, the movie theatre *Le Grand Action*, art schools (*École Boulle*, *École nationale supérieure des Arts Décoratifs*), and *MMI* and *Maison de Fermat* elsewhere in France.

Furthermore, we have been working with many not-for-profit associations that are engaged in mathematics communication, in particular those which are hosted at IHP. We are also in close contact with *Les Maths en Scène*,[v] *Science Ouverte*,[w] which works at popularizing science in high schools from disadvantaged areas, and the *Institut Télémaque*,[x] which offers support to students of modest origins and helps them to gain confidence in their training paths, in particular by having them sponsored by companies.

Challenges

We are a few months away from the opening and we still have many challenges to meet.

---

[u] The city of Paris, the Ile de France region and the French state.

[v] https://lesmathsenscene.fr/.

[w] This name here does not refer to Open Science in the science publishing business. It stands for openness to all types of public. See https://scienceouverte.fr/.

[x] https://www.institut-telemaque.org/.

Fundraising

So that the Maison Poincaré can function properly, we must have a great fundraising strategy with companies and individuals. This implies expanding our network of companies by meeting people in the digital, finance, energy, etc., fields. The IHP endowment fund is in charge of carrying out this fundamental mission, which has unfortunately been jeopardized by the economic crisis induced by the pandemic. We must certainly engage even more in helping them to raise funds.

Another crucial item in the business model for the extended IHP, and in particular the Maison Poincaré, lies in getting revenues from the Perrin building itself. We are convinced, and this has been confirmed by our network that given its location and beauty,[y] there will be some demand to rent seminar rooms, salons, and even the garden of the building. Nevertheless, we still have to create a credible offer to run the promotion of these premises efficiently and rent them profitably.

Inclusivity

It is highly desirable if not mandatory that our exhibitions and events be accessible to visitors with disabilities. Even though we are very much aware of this, and scenographers and graphic designers as well, it is far from being obvious how to adapt exhibits to all kinds of disability. We have been working with students and science communicators with disabilities as advisors, but there is still work to do.

Gender equality is also kind of a challenge in our projects. For instance, it is sometimes very hard to find female speakers for the film club or documentary films. We are also aware that gender equality has so far not been achieved in our temporary exhibition themes. This issue will be addressed in the upcoming programming. On the other hand, in the permanent exhibition, we have chosen to balance the visibility of women with that of men despite their appalling underrepresentation in the STEM community. Even though it will satisfy many people, we know that we may

---

[y] We have not described the upper floors here but they will be spacious and gorgeous.

face criticism for this choice, as is already the case regarding our use of inclusive writing in our communications.

Attractiveness

Dialogue between researchers and the public is another challenge. Indeed, our visiting researchers are on-site but it is not clear whether they will be willing to interact with the public. The language barrier is one of the difficulties to cope with, since the French laymen are hardly comfortable with talking to strangers in English. We must find easy ways and original incentives to stimulate discussions and exchange, be it through organized or informal encounters. This mission will be part of the job of our permanent team of science communicators, provided that we can secure positions.

Last but not least, even though our campus is attended by lots of scientific students and academics, it is much less known by the public. So we must change this and advertise our campus, the institute, our activities, and the Maison Poincaré. The help of the city of Paris is going to be crucial in this matter.

We will consider us successful only if junior and senior high school students, in particular from disadvantaged areas, come in numbers to the Maison Poincaré. To achieve this goal, we must mobilize and involve teachers, teachers' associations, libraries, universities, and school boards. We would consider a minimum of 15,000 students (600 class groups) and 5,000 other visitors per year as a success.

Conclusion

Of course the opening of the Maison Poincaré, one that we all look forward to, will be a crucial step in the project, but is definitely not the last one.

Meanwhile we will have offered virtual visits to our targeted audience and partners, and tested with teachers and their students a number of hands-on exhibits.

It is our intent to continue our collaborative design methods and improve our offerings by taking into account feedback from our public in the long run.

We will also continue to share good practices and circulate our productions within our national network so that more people can benefit from them.

More generally, we have the ambition of becoming a source of inspiration regarding mathematical communication, in partnership with our national and international network.

References

[1] Torossian, C., and Villani, C. (2018). *21 mesures pour l'enseignement des mathématiques*. Rapport, Paris: Ministère de l'éducation nationale, de la jeunesse et des sports.

[2] Kogelman, S. and Warren, J. (1978). *Mind Over Math*. New York: McGraw-Hill Company.

[3] Bishop, A. J. (1998). "Cultural conflicts and social change: Conceptualising the possibilities and limitations of mathematics education". *MEAS Proceedings*. Nottingham, 12–16.

[4] Perrin, J. (1913). *Les atomes*. Paris: Félix Alcan.

[5] Villani, C. and Uzan, J.-P. (2017). *Objets mathématiques*. Paris: CNRS Editions.

[6] Villani, C., Uzan, J.-P., and Moncorgé, V. (2017). *The House of Mathematics*. Paris: Le Cherche-Midi.

Gueudet, G., Bosch, M., Disessa, A., Kwon, O. N., and Verschaffel, L. (2016). *Transitions in Mathematics Education*. Berlin: Springer.

Links

https://www.ihp.fr.
https://patrimoine.ihp.fr/.
Interviews: https://images.math.cnrs.fr/-L-institut-Henri-Poincare-IHP-.html.

https://doi.org/10.1142/9789811253072_0008

Chapter 7

Math Science Communication Without Words: Mathematics Adventure Land, Dresden

Rahel Brugger, Bernhard Ganter, and Andrea Hoffkamp

Erlebnisland Mathematik, Dresden, Germany
service@museen-dresden.de

Mathematics Adventure Land, a well-attended permanent exhibition in a municipal museum in Dresden, Germany, was established in September 2008 as the result of a co-operation between the city's technology museum and the mathematics department of TUD, Dresden's renowned university of technology. We sketch the concept of the exhibition and explain its contribution to science communication.

As of now, Mathematics Adventure Land has existed for over a decade. An exhibition living for so long cannot be static, it must evolve. The organizing team is constantly developing a variety of additional offers to complement the current exhibits. The three authors of this article represent different aspects of this development: Bernhard Ganter, a now retired professor of mathematics, was one of the two founding directors and initiated many of the exhibits himself. In section "Origin, Motivation and Guiding Principles", he will report how the project came about and which ideas were implemented in the process. Rahel Brugger holds a Ph.D. in mathematics and presently undergoes a traineeship in the museum (Volontariat). She reports in section "Education Formats in and Around Mathematics Adventure Land" on the many additional activities that have been initiated and realized. Section "Mathematics Adventure Land and Teacher Training", written by Andrea Hoffkamp, who is a professor of didactics of mathematics and also a scientific director of the exhibition, shows how the exhibition contributes to the teacher training in co-operation with TU Dresden.

Origin, Motivation and Guiding Principles

How it started

In early 2007, a businessman had a plan. He had visited Beutelspacher's museum *Mathematikum* in Giessen, Germany [1], was enthusiastic about it, and wanted to set up a similar exhibition in Dresden. This succeeded, albeit differently than planned, because private financing did not materialize. Financed by funds from the Saxonian state government and the city (and without participation of the disappointed businessman), a permanent exhibition was established in a municipal museum, the Technology Collections Dresden (TSD).

Beutelspacher had recommended two math professors, Volker Nollau and Bernhard Ganter, as experts, and they now had to take over the project. The two brought very different experiences. Nollau was from Dresden and knew his way around. In addition to his work as a stochastics professor, he was politically active and at times even held the office of State Secretary. Without his knowledge of the decision-making processes, no state funding would have come about. Ganter, an algebraist, came to Dresden only after the German reunification and did not know much about the political situation there. But he had some experience in organizing large exhibitions and also in science communication. Both complemented each other without dissonance. Nollau coped with all the pitfalls of filing the application, while Ganter prepared a plan for the exhibits.

The financing commitment was given at the end of December 2007. One condition was to open in 2008, which was the *Year of Mathematics* in Germany. So the exhibition had to be designed and implemented within a few months in phases of feverish work, parallel to the renovation of a part of the building. Figure 1 gives an impression of the situation in March 2008, Figure 2 of that half a year later. Without purchasing numerous replicas from

Figure 1: The condition of the exhibition rooms a few months before the opening.

Figure 2: The polyhedron of triads.

the *Mathematikum*, the exhibition would not have been feasible in such a short time. When it opened in September 2008, about half of all exhibits came from there. But both sides agreed that the Dresden exhibition should not become a copy of the Mathematikum. An original conception had to be developed, with original content and design, and all in the shortest possible time. Those familiar with the magnificent famous museums in Dresden would perhaps assume that a large team was provided for such an exhibition development. But the "Old Masters Collection", the "Green Vault", etc., are *state* institutions. The *municipal* museums are financed much more sparingly. They can only achieve their respectable performance by deploying their staff in a network. The museum director Dr. Lindner was only able to delegate a single research assistant (Dr. Michael Vogt, who later became director of the *Verkehrsmuseum* Dresden) to the development of the mathematics exhibition. He was responsible for all aspects of the implementation. Antje Werner, a freelancing architect, was commissioned to develop the aesthetic design of the exhibition. She worked closely with the other three. Of course, four people could not set up such an exhibition on their own. They had extensive support from the museum staff network. But the core task of the realization lay with these four.

A clear concept

The fact that our initially vague ideas about the exhibition were by no means met with approval, but on the contrary with skepticism and opposition, ultimately turned out to be useful. It forced us to follow a clear, well-founded concept. The superficial intentions were clear: mathematics should be popularized. There had already been many efforts to promote the rather tepid interest in the economically important STEM subjects. We were supposed to make a contribution to this, which was what the funding was for. But how?

There were good reasons to be cautious. The often repeated assurances that mathematics is *important, universal,* of *incomparable beauty,* and even *God's language,* had obviously not been convincing. It was still considered normal and beneficial for one's personal reputation to publicly flirt with one's mathematical stupidity. Even the professional advertizing for the Year of Mathematics, which combined mathematical facts with everyday situations of young people, had an ingratiating rather than convincing effect: the flight parabola means little for a skateboarder's driving experience.

We were therefore determined to largely dispense with explanatory and instructive texts. Obviously the problem is not inadequate information, but the fact that mathematical thinking is not inviting enough. This should be worked on, and verbal assurances were evidently an ineffective tool.

So visualization? Here, too, we were skeptical. Illustrating abstract facts is of great help for mathematical thinking. Good visualizations support working with abstract ideas. But they are not interesting and attractive by themselves. Every university has one or many showcases in which carefully designed mathematical models get dusty and are largely ignored even by mathematicians. Geometric shapes, which are spectacular to look at due to their regularity, occur frequently in architecture and are therefore commonplace, but often without triggering interest in their mathematical background.

But what then? Is it "fun" that attracts visitors? In fact, exhibitions that confront visitors with exciting "phenomena" have had many successes. The *Phänomena* [2], held in Zurich in 1984, delighted an audience of millions and continues to have an impact today. Some of its exhibits have found their way to the Mathematikum and also to our adventure land. But focusing on the phenomenon is not enough to get closer to mathematical thinking. Nevertheless, such exhibitions are an important role model because of their "hands-on" principle: The visitors are activated, taken out of the role of the passive observer, seduced into their own actions, and thereby connected to the contents of the exhibits. They should be offered an *experience.*

The name was born: *Erlebnisland Mathematik (Mathematics Adventure Land)* would be the name of the exhibition.[a] Admittedly a gaudy title. But the name is right, it is understood, and is well received.

So what did we aim for, what is a mathematical experience, especially one that gets by without text? We did not intend to present the intellectual adventure journeys of professional mathematicians. The exhibition is aimed at the general public,[b] at all open-minded visitors. It is designed to invite ordinary people to mathematical thinking. This certainly is not merely a process of information or instruction. It also includes aspects of, e.g., aesthetics, trust, self-confidence and meaning. It was therefore important that the designer created a beautiful, bright, and uniform appearance. And it is crucial that the visitors are not confronted with unverifiable allegations, gradings, or exhortations. Instead, they can decide for themselves whether they find what they see important and interesting. And rather than elaborating on mathematical details (and limiting itself to the simple), the exhibition impresses with attractive exhibits that do not easily reveal all of their secrets.

How can that work? If you want to know, you have to visit the Mathematics Adventure Land in the Technischen Sammlungen Dresden and experience around 100 exhibits. Not everything turned out well, but we are happy with what we have been able to achieve.

Is it really true that the exhibits are not explained? Yes, there is no explanation text at the exhibits, just a few words to trigger an action. But visitors who actively ask for explanations will get them. We also offer

---

[a] The English name does not exactly translate the German one.

[b] "Allgemeinheit" as discussed in the work of Rudolf Wille (1937–2017). Wille propagated a "General Science" that faces the general public. See, e.g., [3] for a quick introduction.

plenty of didactic material on the internet. More on this will be reported in section "Education Formats in and Around Mathematics Adventure Land".

Working theses

Mathematical work usually starts from definitions, which precisely describe the notions that are investigated. When our exhibition concept had to be implemented under great time pressure, we would have loved to have definition-like formulations of our objectives. What exactly is Mathematics Adventure Land supposed to do and how are its goals achieved? But of course we could not expect to find definitive answers to these questions. Even "Mathematics" has, according to Wikipedia, *no generally accepted definition*.

At that time I formulated a few theses that I kept coming back to mind. The formulations are not fully developed; the reader may take them as a basis for their own thought:

1. MATHEMATICS ADVENTURE LAND INVITES THE GENERAL PUBLIC TO PROMOTE MATHEMATICAL THINKING.

   (*Mathematical thinking* in its broadest sense, not restricted to the systematic or deductive, mathematical *thinking*, not skills and knowledge first, the *general public*,[c] not only a selection of "gifted" individuals, *promote* in the sense of encourage, stimulate and strengthen.)

2. MATHEMATICAL THINKING IS FUNDAMENTAL IN MANY WAYS FOR OUR KNOWLEDGE CULTURE AND IS A PREREQUISITE FOR PARTICIPATION.

   (*Fundamental* expresses that it cannot be reduced to simpler principles, *in many ways* emphasizes that a one-time communication process is not sufficient, *our knowledge culture* refers to the present time, *participation* stands in contrast to passive acceptance and can be orienting, productive, or critical. Rational decision-making (and participation in it) often requires proficiency in logical and quantitative argumentation.)

3. MATHEMATICS ADVENTURE LAND AIMS TO IMPROVE ATTITUDES TOWARD MATHEMATICAL THINKING AND CONVEY MATHEMATICAL EXPERIENCES THROUGH THE ENCOUNTER WITH THE EXHIBITS.

---

[c] Compare footnote b.

(Systemic training is a matter of mathematics teaching in schools, not of our exhibition. *Improving attitudes* includes opinions, feelings, and values, *encounter* requires a reaction from the visitor, with a positive memory value.)

4. MATHEMATICS ADVENTURE LAND EMPHASIZES THE DIVERSITY OF MATHEMATICAL THINKING.

However, these theses do not answer the question of what mathematical thinking is. We do not have an answer. But at least we can formulate a property:

5. MATHEMATICS REQUIRES GOAL-ORIENTED, COMMUNICATIVELY JUSTIFYING, OFTEN SYMBOLIC THINKING IN ORDER TO EXPLORE POTENTIAL REALITY.

Communication thus is essential for mathematics. Mathematical results require proofs. They must convincingly be justified in communication with other mathematicians. The author of these lines believes that it would be beneficial for the mathematical community to find and enhance communication with the general public as well, in a two-way, not merely informative, process. So far however, Mathematics Adventure Land supports such a reciprocal science communication process only very indirectly.

Two examples

Unfortunately, the rules outlined above do not help in *inventing* exhibits. And even if an exhibit design fits the theses, there is no guarantee that visitors will find it attractive. For exhibition organizers, it is heartwarming when an exhibit really works. As successful examples, I briefly present two (of the about 100) objects in the exhibition.

The *Polyhedron of Triads* is a tower of five regular octahedra of the same size, stacked on top of each other (see Figure 2). It illustrates aspects of elementary musical harmony. The nodes of the polyhedron correspond to the notes of the chromatic scale and are labelled accordingly. Its faces, all being equilateral triangles, correspond to the harmonic triads (except for the top and bottom face), and all such triads do occur. More precisely, triangles pointing up represent major triads, while those pointing down represent minor triads. Other patterns, like the circle of fifths, are easily discovered.

The Polyhedron of Triads had been designed (by the author) for a much earlier exhibition, at that time not as a physical exhibit, but as a paper cut-out sheet that was distributed in large numbers and later published as a PDF. Even years later there was still interest in this little sculpture, and so it was only natural to set up a more spectacular, playable version for Mathematics Adventure Land. A professional model construction company created a man-high sculpture made of metal and acrylic glass, the vertices of which actually produce the specified tones when touched. That was expensive, but it was worth it. The polyhedron is one of the highlights of the exhibition and is constantly being used.

The visitors are drawn into the exhibit, no matter if they are interested in music theory or not. You touch it and there are tones, interesting! You try to play a melody, but find that this is difficult because of the unusual geometry.[d]

The sign on the exhibit says it is about triads, and indeed the triangles sound good! One begins to understand that there is a geometry of triads, and that this geometry is clearly structured, but non-trivial, and perhaps useful. In this sense, the exhibit represents an invitation to the general public to use mathematical thinking, here in understanding the elementary rules of musical harmony.

Musicians do not need the polyhedron to understand their rules. On closer inspection, however, they too will discover something interesting. The top octahedron is just a repetition of the bottom one, but twisted. The sculpture is therefore not to be understood "modulo octave", but as a section from the potentially infinite chromatic scale. If you were to identify tones at octave intervals, a torus structure would result, not a tower. And so on, and so on: the exhibit makes it possible to explore mathematical structures in music, to delve into them without reaching an end. Occasionally I have met music teachers there, giving their students long spontaneous lectures at the polyhedron.

---

[d] The model builders were smart enough to set up the electronics in such a way that if played too wildly, it pauses and remains silent for a few seconds.

My second example is a much more modest exhibit, called *Klarner's Theorem*.[e] On a table there is a 10 × 10 grid of equally sized squares, together with 1 × 4 tiles, 25 of them. The exhibit text asks you to lay out the grid with the tiles, which seems possible because of 10 × 10 = 25 × 1 × 4. At the same time, a reference is made to the help button, which should give an indication of how to prove the impossibility of the task.

If you press the help button, however, you will not get an explanation in words. Instead, the cells of the 10 × 10 grid are colored with four colors in such a way that each 1 × 4 tile covers four cells of different colors, no matter how it is positioned on the grid. It only takes a little flash of inspiration now to understand the impossibility of the task. So the exhibit guides the visitor to find an actual proof, proving that something is impossible!

Another exhibit, named *Conway's Cube*,[f] is a puzzle requiring you to fill a 5 × 5 × 5 cube with 29 bricks of size 1 × 2 × 2, plus three of size 1 × 1 × 3. This puzzle is solvable, but not easy, in particular since no hints are given. A solution is much easier to find when using a coloring argument similar to the one used for *Klarner's Theorem*. Yes, exhibits are interconnected. Not systematically, but some share common secrets. Discovering these is the pleasure of our mathematically-knowledgeable visitors.

---

[e] Named after the US mathematician David A. Klarner, 1940–1999.
[f] After the British mathematician John H. Conway, 1937–2020.

Education Formats in and Around Mathematics Adventure Land

Exploring the exhibits

The exhibits in Mathematics Adventure Land can be explored without prior knowledge, additional help, or explanation. However, there are situations in which we do use words (written and spoken), contrary to what the title of this article suggests. Sometimes words become necessary to get visitors — especially adults — used to the specifics of the exhibition. Words may also be required if an exhibit has been successful in getting a visitor so interested in a particular topic that she wants more information than the exhibit itself can provide. Especially with school visits, words might be used to give structure and context, and connect the exhibits to things learned in school.

Like most science centers, Mathematics Adventure Land works with so-called explainers, with whom the visitors can talk about the exhibits if they wish. The role of these explainers is very different from old-fashioned museum guides and the name might be a bit misleading: The way they interact with visitors is focused on dialogue rather than on transferring information from the explainer to the visitor. They might answer questions, ask questions, maybe hint to some interesting observation without giving away everything and act as a role model by being curious and not knowing everything themselves.

In order to get both visitors and explainers more used to this form of interaction, we started the format *Maths Live* where the joint exploration of one exhibit is announced for a fixed time. Typically, two to ten visitors then gather around that exhibit and, together with the explainer, try to find out what it is about. The explainers prepare for these explorations and roughly follow a scheme (some context, experiencing the exhibit, looking for explanations, connections to other exhibits and applications, follow up questions), but still every *Maths Live* is different and the main task for the explainers is to let the visitors make the key observations themselves.

Since recently, we also use the app *Actionbound* as an individual, playful, and flexible tool that can help visitors explore exhibits. *Actionbound* is an app to create all kinds of rallies. It uses gamification elements like points and immediate acoustic feedback, and gives the creator lots of options for different question formats, working with pictures, sound, and individualising the rally depending on the answers a player gave.

The first *Actionbound* rally was created by Elise Stroetmann as part of her thesis [4,5]. Working as an explainer in the exhibition, she had a lot of experience about how the visitors engage with the different exhibits. For her rally, she selected those exhibits where, in her observation, visitors have the most difficulty getting to the mathematical content on their own. She also wanted to cover many different areas of mathematics to give a good impression of how big the variety of mathematical ideas behind the exhibits was. In her rally, visitors who wanted to dig deeper could now explore exhibits that they choose, answer questions about them and get information about the mathematical background on demand.

As a teaching student, Elise had pupils from 11 years up in mind as the main target group for her rally. In fact, most visitors of Mathematics Adventure Land are pupils on a school trip and families with children. However, the exhibition is suitable for adults, too. The other two *Actionbound* rallies are therefore mainly targeted at adults and teenagers. One is about exhibits that have some connection with an open mathematical problem [6]; another one connects some exhibits about the intersection between mathematics and music [7].

Many exhibits have a connection to school curriculum and therefore the potential to let pupils explore and experience the school subject matter in a lively manner. However, it is surprisingly difficult to make sure that all pupils notice certain things about (or even visit) a specific exhibit: Most importantly, there is a risk that the fun will be lost if you replace playing and exploring by more directed activities. In addition, there are practical difficulties addressing a big group in the exhibition since there often is quite some noise and the exhibits are not designed to be explored by more than three or four people at the same time.

So knowing that this is a difficult task, we are at the moment designing material that teachers can download from our website and use if they want to address a certain mathematical topic in their visit. In his thesis, Tom Borrack worked out what he called "research notebooks", which allow pupils to explore certain related exhibits individually [10]. In a slightly different format, we design a shorter version of the research notebook, which we call a "research slip", and combine it with a worksheet to be used in class after the visit to the exhibition. The research notebooks and the research slips are mainly about observing and experimenting with exhibits, while the worksheet addresses questions that arise naturally when doing so but need some time and a calm environment to be answered. These questions might be something like "Why is my mirror image turning upside down when I rotate these mirrors but not with the other ones?" or "Why does this hologram appear here?" or "How are the lengths of these pendulums chosen to give this stunning effect?"

Workshops

Many school classes that visit Mathematics Adventure Land also book a workshop. These workshops are typically 90 minutes long. Some of them partly take place in the exhibition but as described above, it is a challenge to experience the exhibits in a group. So instead, we try to find other activities that many people can do at the same time and have an Adventure Land vibe in the sense that they are hands-on, individual, and playful.

For example, in our workshop about Seifert surfaces, we start in the exhibition where everybody can crawl through our giant knot in order to introduce the topic. Later, instead of talking about the Möbius strip in the exhibition, the pupils make their own one and experiment with it. Similarly, our workshop about tilings focuses on creating one's own individual tilings whereas the many tilings in the exhibitions are better explored after the workshop or in another visit.

These crafting activities most often take place in our new workshop areal called *MACHwerk* (makers' workshop), which is also a maker space with, for example, a laser cutter, a kiln, and a 3D printer.

We use these devices, for example, in our science camp on ornaments. In 5 days during the school holidays, kids and teenagers from 10 to 14 years old make all kinds of different ornaments (see Figure 3) working

Figure 3:   Impressions from the ornament camp. The kangaroo tiling is from the exhibition.

with paper, a ruler and compass, the laser cutter, clay, and the *iOrnament* app [8]. Doing so, mathematical questions about symmetries and tilings arise very naturally. We address these questions when they appear, but also in riddles and our daily warm ups, where we, for example, construct ornaments with our bodies.

Up to now, our workshops focus on children and teenagers but we would like to open this up for adults, too. As an open format for everyone, we have mathematical crafting activities that are not workshops but where you can drop in and drop out at any time. This has been a tradition for many years in the context of special family Sundays and, in addition, we are planning to establish a smaller version of this on a more regular basis every weekend.

During the last year, we made small videos for our YouTube channel with some content from our workshops and family Sundays. At the moment, we are also experimenting with live online versions of our workshops.

Mathematics Adventure Land and Teacher Training

The concept described in section "Origin, Motivation and Guiding Principles" focuses on the experiential character, the activation, the aesthetics, and the promotion of self-confidence, in other words, a positive active experience with mathematics. In the working theses, communicating and reasoning are considered essential and characterizing for mathematics. These aspects also play a central role in didactics and thus also in teacher training. Even though it is often complained that in school the procedural elements of mathematics are too often in the center, the work of Winter [16] on discovery learning in mathematics education was formative in teacher education in German-speaking countries. In addition, there is now much empirical work in the last two decades on teachers' mathematical beliefs and attitudes that are effective in teaching, both in a favorable and unfavorable sense, see Furinghetti and Pehkonen [13], or Staub and Stern [17]. However, cognitive activation, en-active approaches to mathematical content, and the promotion of mathematical communication and reasoning are also central to teacher education. Mathematics Adventure

Land, with its specific conception, therefore offers a great potential to have a positive impact on teacher education as well.

Teacher training within Mathematics Adventure Land

Teacher education in Germany differs from that of most other countries. If you want to become a teacher, you usually decide on this goal before you start your studies and begin a teacher training program that includes two different subjects, didactics and educational science and pedagogy. Particularly the teaching students for secondary schools take some lectures together with the bachelor students in mathematics. But some lectures are also offered specifically for teaching students.

The training of mathematics teachers is a central task of the Faculty of Mathematics. We are facing particular challenges in guaranteeing the high level of training, conveying content in a lively and motivating manner, and later offering new teachers sustainable and supportive supervision.

Since student teachers with two subjects, didactics and educational sciences, have to deal with much more content in their studies than bachelor students with one subject, the specialized training in mathematics is naturally more limited. Mathematics Adventure Land has an almost inexhaustible potential here, not only because of the way mathematics is presented, but also because of the breadth and diversity, as well as the intra- and extra-mathematical interconnections of the exhibits. In this sense, the exhibits can complement the content of specialized education and at the same time enhance teaching skills.

Over the past years, Mathematics Adventure Land has increasingly developed into a museum that is systematically integrated into the teacher training at university. On the one hand, the student assistants in the museum are largely recruited from the teaching students, who gain valuable experiences of mathematics as a living and dynamic science. This enables them to enter their later careers with a broader spectrum of content and methods. On the other hand, seminars with various concepts for the mathematics education of teacher students are carried out on the university side in cooperation with Mathematics Adventure Land. The students carry out mathematical research projects first and present their results in the museum — sometimes in so-called "mathematics shows". Particularly successful contributions are incorporated into the museum's educational program and thus have an impact on the museum's development. Some of these contributions are already mentioned in section "Education Formats in and Around Mathematics Adventure Land". The supervision of such seminars or other contributions is resource-intensive, but rewarding, as they have a lasting positive impact on the image and beliefs of mathematics among teaching students.

Mathematics Adventure Land as a place of didactic research and as a didactic experimental field

Inquiry-based learning and educational design research have become essential quality features of teacher education [15]. The concept of inquiry-based learning in teacher education is meant to mediate not only between education and science, but also between science and profession whereas design research has an innovative character and aims at the development of learning activities in a close theory-practice relationship [14].

At Mathematics Adventure Land, teaching students to research, develop, implement, and evaluate innovative formats and workshops for pupils at various grade levels as part of their final research theses. Some of these contributions are already mentioned in section "Education Formats in and Around Mathematics Adventure Land", e.g., the design of research notebooks for school class visits [10], the ornament science camp [11], and the design of *Actionbound* rallies [5]. However, the range of final theses also includes topics on mathematics and music [9] or the comparison of reasoning in mathematics and in German language, i.e., the comparison of communicating within the different subject cultures [12].

The supervision of scientific theses is usually carried out by a member of the mathematics didactics working group and a researcher in the mathematics faculty. In this respect, the joint activity has an impact on the university by creating a particularly fruitful network within the Faculty of Mathematics, which has a positive effect on teacher education as a whole. Mathematics Adventure Land provides the resources and creates the organizational framework for this. At the same time, the museum's educational program is expanded in a theoretically-based way.

The conscious and systematic integration of Mathematics Adventure Land into teacher training has the overall chance to leave a positive effect on school education by conveying an authentic image of mathematics in the sense of the theses mentioned at the beginning, but also by promoting exploratory and active learning of intra- and extracurricular contents for all age groups.

References

[1] https://www.mathematikum.de/.

[2] https://www.phaenomena.ch/.

[3] Wille, R. (2005). "Allgemeine Wissenschaft und transdisziplinäre Methodologie". In Method(olog)ische Fragen der Inter- und Transdisziplinarität. *TaTup* Vol. 14(2), pp. 57–62.

[4] Stroetmann, E. (2020). *Konzeption einer Rallye mit explorativem Charakter durch das Erlebnisland Mathematik*. Staatsexamensarbeit Fachdidaktik Mathematik, Lehramt an Gymnasien, Technische Universität Dresden.

[5] https://actionbound.com/bound/favouriteexhibits.

[6] https://actionbound.com/bound/beautifulproblems.

[7] https://actionbound.com/bound/mathsandmusic.

[8] https://science-to-touch.com/en/iOrnament.html.

[9] Barnstorf, M. (2018). *Mathematik und Musik — Ein fächerübergreifendes Projekt zum Thema musikalische Stimmungen und deren mathematische Beschreibungen in der Sekundarstufe I*. Staatsexamensarbeit Fachdidaktik Mathematik, Lehramt an Oberschulen, Technische Universität Dresden.

[10] Borrack, T. (2020). *Geometrie erleben — ein museumspädagogisches Angebot im Erlebnisland Mathematik*. Staatsexamensarbeit Fachdidaktik Mathematik, Lehramt an Gymnasien, Technische Universität Dresden.

[11] Hederer, S.-A. (2018). *Didaktische Aufbereitung einer Projektwoche zum Thema Ornamente für die Altersstufe 12 bis 14*. Staatsexamensarbeit Fachdidaktik Mathematik, Lehramt an Gymnasien, Technische Universität Dresden.

[12] Fischer, T. (2018). *2 + 2 = 5 — Argumentieren in den Fächern Deutsch und Mathematik am Außerschulischen Lernort*. Staatsexamensarbeit Fachdidaktik Mathematik, Lehramt an Gymnasien, Technische Universität Dresden.

[13] Furinghetti, F. and Pehkonen, E. K. (2002). "Rethinking characterizations of beliefs". In G. C. Leder, E. Pehkonen, and G. Törner (eds.), *Beliefs. A Hidden*

Variable in Mathematics Education? (pp. 39–57). Dordrecht: Kluwer Acad. Publ.

[14] Gravemeijer, K. P. E. and Cobb, P. (2006). "Design research from a learning design perspective". In J. Akker, K. Gravemeijer, S. McKenney, & N. Nieveen (eds.), *Educational Design Research.* (pp. 45–85). London: Taylor and Francis Ltd.

[15] Saunders C., Gess C. and Lehmann M. (2020). "Forschendes Lernen im Lehramt". In C. Wulf, S. Haberstroh and M. Petersen, (eds.), *Forschendes Lernen.* Wiesbaden, VS: Springer.

[16] Winter, H. (1989). *Entdeckendes Lernen im Mathematikunterricht.* Braunschweig: Vieweg.

[17] Staub, F. C. and Stern, E. (2002). "The nature of teachers' pedagogical content beliefs matters for students' achievement gains: Quasi-experimental evidence from elementary mathematics". *Journal of Educational Psychology*, 94(2), 344–355.

© 2023 World Scientific Publishing Co. Pte. Ltd.
https://doi.org/10.1142/9789811253072_0009

Chapter 8

The IMAGINARY Journey to Open Mathematics Engagement: History and Current Projects

Eric Londaits*,‡, Andreas Matt*,§, Antonia S.J.S. Mey*,†,‖, Daniel Ramos*,¶, Christian Stussak*,**, and Bianca Violet*,††

*IMAGINARY, Germany
†EaStCHEM School of Chemistry, University of Edinburgh, Edinburgh, UK
‡eric.londaits@imaginary.org
§andreas.matt@imaginary.org
‖antonia.mey@imaginary.org
¶daniel.ramos@imaginary.org
**christian.stussak@imaginary.org
††bianca.violet@imaginary.org

In this chapter, we will take you on a journey through mathematics exhibitions and other mathematics public engagement activities our nonprofit organization IMAGINARY has brought to life. This journey started in 2007 and was made possible through many collaborations and partnerships to get us to where we are in 2021, with more than 100 countries having seen our exhibitions, workshops, or other activities. The idea behind this article is to highlight key decisions we took on our journey, some personal insights, and to introduce our open and participative way to engage the general public with mathematical sciences.

In the first part, we will describe how the IMAGINARY[a] project evolved over the last 14 years and how we have focused on refining our didactical approaches we deem appropriate for public engagement. This makes us a strong advocate for non-formal mathematical education. We have gathered our experience in creating material for outside-of-school workshops and help teachers prepare modern mathematical topics for the classroom, as well as other science educators. Another aspect that we took seriously and, as the project grew, had to dynamically adjust, was our approaches to customizing material for different demographics. The focus was both on the cultural context of different regions in the world, as well as the inclusion of groups that typically would not be exposed to mathematics exhibitions or workshops. This would include empowering girls and women as well as families from poorer socioeconomic backgrounds. In the process we revised our policies, balancing a strong commitment to free and open licenses, with our paid services covering our work. Our network has grown beyond the academic world of mathematicians and now includes different companies, artists, museums, and cultural mediators.

In the second part we will walk through a few of our latest projects that expose what is IMAGINARY nowadays. The highlights are the two exhibitions: *La La Lab — the Mathematics of Music* (2019) and *I AM A.I. — explaining artificial intelligence* (2020). We invite the reader to

---

[a] http://about.imaginary.org/.

follow IMAGINARY on the next steps of its mathematics journey by providing a glimpse into a couple of currently ongoing projects we are involved in: the *International Day of Mathematics* (2020), our new *10 Minute Museum* on the mathematics of the Climate Crisis (2021), and our exploration into interactive children story books through *Mathina* (2021).

Part I: Evolution of IMAGINARY

And it all began in Oberwolfach: The traveling exhibition (2007–2012)

In 2007, the Mathematisches Forschungsinstitut Oberwolfach (MFO), under the direction of Gert-Martin Greuel decided to take part in the German Year of Mathematics 2008, promoted by the German Ministry of Sciences and Education. The original idea, to open a new mathematics museum in the German Black Forest, was postponed and replaced by the idea to develop a traveling exhibition to be shown in 12 cities throughout Germany, one month at a time. This way we gained experience through traveling exhibitions before opening a new museum and this decision, by the way, has proven to be a key step also for other future museums. The traveling exhibition *IMAGINARY — through the eyes of mathematics* focuses on the visual side of mathematics, with algebraic and differential geometry as two core topics. During the organizational phase of the project, we invited mathematicians to participate and curate a compact exhibition (see Figure 1(a)) of five interactive software-based exhibits, one film station, one gallery of 3D-printed sculptures (see Figure 1(c)), and one picture gallery (see Figure 1(b)) [3]. The exhibition was — and it still is — on display in several countries and is a visually stunning experience consisting of very colorful images and a unique interplay of formulas and its resulting forms. It paved the way for how IMAGINARY deals with visitors: they are invited to create, invent, and enjoy. This was done through the creation of new algebraic surfaces. They were invited to explore the exhibit and invent new designs with it. Lastly, visitors were encouraged to experience and enjoy new (or uncommon/ unknown) technologies such as 3D glasses, large touch screen interfaces, or 3D prints. It should be noted that in 2007 and 2008, 3D printing and touch screens were very new technologies, and not yet widely used in exhibitions; for example, it was commonly required to give an introduction on how to use the touch screen functionality to interact with the exhibits. All this led to visitors being challenged with complex mathematics and exciting open questions in current mathematical research topics.

(a)

(b)

(c)

Figure 1: (a) IMAGINARY exhibition in Hanover, Germany, 2010. Image credit: IMAGINARY. (b) IMAGINARY exhibition in Istanbul, Turkey, 2014. Image credit: Christian Stussak. (c) IMAGINARY exhibition in Seoul, South Korea, 2014. Image credit: David Grünberg.

Another key aspect for us was to listen to visitors' feedback and constantly adapt or enhance the exhibits. We have always considered our work and exhibits as dynamic entities and kept on changing exhibits or exhibit interfaces, even while the traveling exhibition was on display. One key decision, for example, was to add a printer to the exhibition. Visitors were now able to take their creations, an algebraic surface with its polynomial equation, home or also leave a copy of it at the exhibition. These user-generated galleries were extremely attractive, not only for the visitors creating them but also for others to see what fellow visitors had left. Due to its popularity, we even created an online gallery and competition for algebraic surfaces with several thousand participants. These competitions were run in collaboration with major newspaper media in Germany [14].

We like affirming visitors' requests at IMAGINARY. "Yes, you can copy the exhibition and reproduce it. Yes, it is all free." was the answer to the visitors' questions, which eventually led to our open-source approach. We first shared digital data, for example, to reprint the images of the exhibition, or the use of the software in schools, and then prepared downloadable packages, ultimately leading to the IMAGINARY online platform that we describe below.

We want to highlight one exhibit in particular: *SURFER*,[b] a real-time raytracing program displaying solutions to polynomial equations as algebraic surfaces (see Figure 2(a)). It somehow embodies many of our core concepts within one exhibit: it aims to provide an intuitive connection to a complex pure mathematics topic with open research questions, it is beautiful and creative, and an amazing tool to create mathematical images [4,5]. It has been used in museums, schools, by artists, fashion designers (see Figure 2(c)) and even by a chef to create mathematical 5-star cuisine (see Figure 2(b)).

In 2010, we opened the Mathematik Mineralien (MiMa) — a museum for minerals and mathematics in Oberwolfach, just a short walk from the MFO, with the key exhibits of our traveling exhibition and some new additions. The traveling exhibition itself started touring outside of Germany as of 2009 and until 2013 was shown in many different, but mainly European countries. Spain was the first country to show a countrywide IMAGINARY exhibition, organized by the Spanish Mathematical Society, RSME [7].

---

[b] http://imaginary.org/de/program/surfer.

(a)

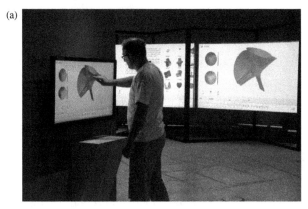

(b)

(c)

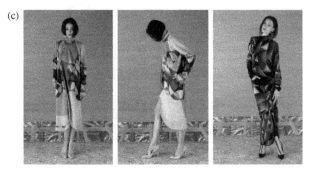

Figure 2: (a) SURFER exhibit at IMAGINARY exhibition in Kaohsiung, Taiwan, 2015. Image credit: Bianca Violet. (b) SURFER algebraic surface 5 star cuisine. Image credit: IMAGINARY. (c) SURFER, Fashion, Ljubljana. Image credit: Draž Fashion.

The open-source exhibition platform: Facilitating the hosting and spread of open source exhibits (2013–2016)

In 2013, IMAGINARY experienced its first significant transformation from being a single exhibition project to being a multi-exhibition project, an online platform, a community of users and partners, and a stable working team. This opened a period of even further expansion, a consolidation of the project's open-source philosophy, and an exploration of new communication formats beyond the exhibitions.

IMAGINARY evolved into an open platform[c] for the communication of mathematics in English, German, Spanish, French, Turkish, and Korean. It represented a new way of communicating science: collaborative [13], global [11], free, and close to research. Everybody was invited to join, submit exhibits, or download them and organize local and individual exhibitions [8–10]. This led to the creation of an international network to connect math communicators.

With the launch of the platform, a second major exhibition was presented: *Mathematics of Planet Earth* (MPE). Following our participative approach, and the positive feedback from doing competitions, we decided to create the whole exhibition via an international call for exhibits. The exhibition shows how mathematics can help to understand phenomena of our planet and was launched at the UNESCO headquarters in Paris [18]. To date, the MPE exhibition, or parts thereof, has been shown in 17 countries around the world. We had a second competition round of exhibition entries which were shown and awarded at the Imperial College London in 2017 (see Figure 3) [2].

With additional exhibition content and the open platform, an ever-growing list of IMAGINARY exhibitions, workshops and training events happened and the core team of IMAGINARY expanded. Particularly, a handful of freelancers available for organizing regional exhibitions made a lot of the work of exhibition setup, training of exhibition guides, etc., possible.

The exhibitions began to spread and IMAGINARY supported many traveling exhibitions, which were produced and organized by local

---

[c] http://imaginary.org/.

Figure 3: Mathematics of Planet Earth exhibition in London, UK, 2015. Image credit: Christian Stussak.

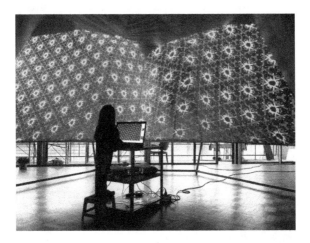

Figure 4: Traveling exhibition "IMAGINARY — un viaje por la matemática" in San José, Uruguay, 2016. Image credit: IMAGINARY Uruguay.

teams: Spain (2010–now), Russia (2012–2013), Uruguay (2015–now) (see Figure 4), Turkey (2015–now), Israel (2015–2016), France (2015–now), Belgium (2015–2016), Netherlands (2016–2017), Taiwan (2016–now). We provided the software and data for the exhibits and additionally encouraged to include exhibits of local authors. For some of the exhibitions, we led the organization of the exhibitions, curated the exhibits, installed the software on the local machines, etc. For others, the local team took over most or all of the organizing. Often, we helped with training the local guides, held workshops on-site, provided infrastructure for online competitions, and helped out wherever possible. Some exhibitions are still traveling around the countries where a unique version of an IMAGINARY exhibition was created. And some have had a huge impact on how mathematics is perceived, see for example the results of a several-year-long traveling exhibition in Uruguay [1].

There were many more temporary exhibitions in single locations. For an overview of all events that took place between 2007 and May 2021 have a look at our event map (see Figure 5).

There were a few exhibitions that are noteworthy and purpose-built by IMAGINARY: at the International Congress for Mathematicians (ICM) in 2014 in Seoul, South Korea, the local research institute NIMS presented the biggest exhibition so far with 12 interactive programs presented on large touch screen panels, a printed picture gallery, a digital picture and film gallery on the latest mathematical animations and visualizations, and a sculpture gallery with 3D prints. It welcomed more than 10,000 visitors in 9 days, among them many local visitors and schools. In some countries, IMAGINARY exhibitions have been the first (or among the very first) mathematics exhibitions ever, for example in Panama or Cape Verde.

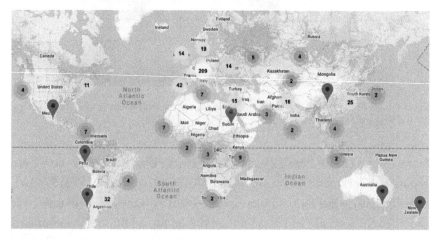

Figure 5: Event map of IMAGINARY activities (between 2007 and May 2021).

IMAGINARY has been active on the African continent since 2014 in collaboration with UNESCO, the African Institute of Mathematical Sciences (AIMS), and the Next Einstein Initiative. Our presence has included training workshops in mathematics communication and our exhibitions. We have organized events in countries such as Liberia, Senegal, Democratic Republic of Congo, Cape Verde, Tanzania, Rwanda, South Africa, and Algeria and participated in large conferences such as the Next Einstein Forum Global Gatherings. We also showed some of our latest projects for the first time on the African continent, for example, the augmented reality project Holo-Math,[d] in collaboration with Institut Henri Poincaré and other partners (see Figures 6(a) and 6(b)).

Many exhibitions were permanently installed in science centers and museums, and some major museums also added individual exhibits, for instance the Deutsches Museum in Munich, Germany, the MoMath in New York, USA, or the Ramanujan Math Park outside Bangalore in India.

As IMAGINARY grew, its importance in math communication was also recognized more and more, resulting in numerous awards including the "Medienpreis Mathematik" by the Deutsche Mathematiker-Vereinigung (DMV) in 2013, which was awarded to Gert-Martin Greuel and Andreas Matt for their work on the project IMAGINARY. The prize is awarded for outstanding contributions to the communication and popularization of mathematics. In 2020, the open-source exhibition concept that permeates IMAGINARY's philosophy was awarded the ECSITE Mariano Gago award for sustainable success.

Other than offering guided school visits to our exhibition, in 2013 we took our first step toward engaging more directly with schools and prepared a special package to fit the requirements and needs of students and educators: the IMAGINARY-Entdeckerbox (discovery box), an open-source package with 25 interactive and hands-on mathematical contents for schools was created [12].

As a result of the many international partners and collaboration projects we had, which we have always enjoyed, we naturally started to work more and more on networking, providing services to the community,

---

[d] https://holo-math.org/.

Figure 6: (a) Holo-Math at the Palais de la découverte in Paris, France, 2017. Image credit: Camille Cier. (b) Presidents Paul Kagame (Rwanda) and Macky Sall (Senegal) testing the Holo-Math project. Kigali, Rwanda, 2018.

connecting partners, sharing ideas and in 2016 also organized our first IMAGINARY conference (see next section).

Internationality, innovations, and independence (2016–2020)

In 2016, a second transformation took place when IMAGINARY became a German nonprofit organization, independent of the MFO (which remained a shareholder). With the initial financial support of the Leibniz

Association, IMAGINARY started as an individual organization with the goal of self-sustainability, offering services to museums and foundations, and creating exhibitions and other products on demand. In this last period between 2016–2020, IMAGINARY has developed two full-fledged new exhibitions. At the same time, it has also diversified its projects to include workshops with school audiences, online outreach projects, conferences and events, IT services to museums, EU projects on educational innovation, and many others.

In principle, our journey continued keeping the open-source approach alive but adding a new services project-line, where partners asked us to devise projects for them. One of the first exciting new projects was the mathematical shopping center Mathematikon, which opened in 2016 in Heidelberg, Germany [21]. Together with mathematicians from 14 countries, we created the mathematical content and were also responsible for the exhibition design. A highlight was the 84-inch touch screen, which was mounted vertically and offered two different math games at the same time at different heights (see Figures 7(a) and 7(b)). A mathematical image gallery with 13 big format prints and explanations was displayed in the general public area and the parking garage, providing unusual insights into different fields of mathematics. The conveyor belts at the cashiers in the supermarket invited you to playfully think about classic and modern mathematics (see Figure 7(c)). You could experiment yourself using the items you want to

(a)   (b)   (c)

Figure 7: (a)–(c) Mathematikon shopping center in Heidelberg, Germany, 2016. Image credit: Bianca Violet.

buy — juice packages, bottles, oranges, eggs… anything. Even the bathrooms had mirrors with mathematical riddles projected onto them.

In collaboration with two Berlin-based collaborative research centers (CRC DGD and CRC STM), IMAGINARY organized a math-art contest for new creative ideas and their realizations based on modern mathematical concepts currently worked on in the two involved research centers. By combining mathematics with art and design, the power, the fun, and the beauty of this interdisciplinary connection was shown. The project included mathematical and creative input for the participating artists, a public presentation of resulting artistic ideas and a final exhibition of the created artworks. The organizers and researchers were in close contact with all teams during the production process of the Math Creations. All artworks were published under a creative commons license and a professional film crew led by Ekaterina Eremenko (see Chapter 12), who recorded all three events as well as the realization process of the artworks and produced a feature film, starring the participants and their Math Creations. This not only disseminates the scientific results of the two collaborative research centers to a wider community, it hopefully inspires more researchers to reach out and connect with other disciplines [19].

The first IMAGINARY Conference on Open and Collaborative Communication of Mathematical Research took place in 2016 in Berlin, Germany (IC16). It was planned to be organized bi-annually, and the second one (IC18) took place in 2018 in Montevideo, Uruguay. The conferences are interdisciplinary gatherings for mathematicians, communicators and interested professionals to discuss and work together on current issues of communication and knowledge transfer in mathematics. In contrast to a classical conference, they emphasize a participative and collaborative approach to advance new ideas, bring forward concrete results, and help shape the future of mathematics communication. Several projects initiated or resulted from these workshops, such as Holo-Math (see above) or Polypad.[e] The third conference needed to be postponed due to the COVID-19 pandemic situation and will now take place virtually in 2021.

---

[e] https://mathigon.org/polypad.

Part 2: Current Exhibitions and New Projects

La La Lab — Exploring our creativity

La La Lab — *the Mathematics of Music*,[f] was our third major exhibition, after *through the eyes of mathematics* and *Mathematics of Planet Earth*. We developed it in 2018–2019, when IMAGINARY was already an independent entity. This time we had a development contract, a larger budget, a dedicated team of our staff, and a global vision for the content and the story we wanted to tell. In the process, we learnt many new aspects of exhibition-making, some of which may be common knowledge for museums and exhibition agencies, but we also found many aspects that would be particular to some of our characteristics: being quite close to the academic world, and having an open licensing and developing philosophy. We will describe here our internal development process, explaining our priorities and guidelines. We hope that this rather intimate description can be an inspiration for other exhibition projects in other associations and math museums worldwide. The reader interested in a more detailed and exhaustive description of the exhibits featured in *La La Lab*, can refer to [15] or to the exhibition booklet [6].

La La Lab came as a request from the Heidelberg Laureate Forum Foundation (HLFF) to be on display at the Mathematics and Informatics Station (MAINS) showroom in Heidelberg, Germany during 2019. The subject of Mathematics and Music was a shared decision with HLFF. Since our first exhibition was devoted to abstract mathematics and the second one to applications of mathematics (for the Earth sciences), entering the domain of mathematics and arts allowed us to broadcast the message of the ubiquity of mathematics, and also provided a point of contact with an audience initially more interested in music than math.

Unlike our previous exhibitions, where external participants submitted modules in a contest and we selected and curated the available options, with *La La Lab* we had the opportunity to decide the content beforehand and to develop many of these modules with a storyline in mind. We did open a call for participation on our website, and some of the user-generated

---

[f] http://lalalab.imaginary.org/.

exhibits were included in the exhibition, but the main emphasis for *La La Lab* was on exhibits we developed in-house. Our initial starting point was to do internal research to start shaping the concrete topics to include, and the type of content we wanted. We designated a curator from within our team but since we had no person in our team with advanced musical studies, an early task was interviewing and gathering a scientific board, to provide scientific rigor for the storyline, to obtain some exhibits from them, and to get overall counsel.

Early on, we set some guidelines for the exhibition concerning different aspects. About the content, we had some essential topics from classical music theory that should not be omitted, but we also wanted to have some cutting edge research. From a pedagogical perspective, the exhibition should offer the ability to learn key ideas, but we also wanted to give some playful and light experiences. The exhibition should also mix and balance science and art, so in addition to accurate math, we had to include real musical pieces, with a variety of styles, and also some artworks such as images or sculptures. Finally, we also set some goals on the design, to make the space immersive and with some unity of style.

From the content side, we identified from classical music theory some indispensable topics to be covered: acoustics and physics of sound, the creation of musical scales and tuning, the theory of harmony and chords, rhythm, and composition. Although other topics could have been also valid, our selection constituted a storyline for the exhibition, building music from scratch: In order to have music, the first thing you need is sound and some instrument that makes it. Then you need to select a set of pitches to tune your instrument with a musical scale. Then you explore harmony and the combinations of notes that match together. You need to add rhythm to impel the music forward in the time dimension, and finally, you put it all together as a composer to create a musical piece. Crucially for the message of the exhibition, all the steps involved in creating music were supported by one form or another of mathematical theory. This constructivist path could be seen in the exhibition, but the exhibition was far from being linear. For instance, with *The Spectrum of Sound*, visitors could explore acoustics and Fourier theory within a two screen exhibit: on the first touch screen there was a spectrum analyzer attached to a microphone, on the second one a graphical synthesizer was shown together with a speaker. The exhibit *Pink Trombone* offered a mathematical model of the human voice that the visitors could control and hear while graphically manipulating an image of the vocal tract with several fingers via the touch display. With the *NSynth* exhibit, visitors could use a synthesizer and create new sounds by finding an instrument consisting of sounded somewhere between two (or up to four) examples, which could be chosen from a library, such as a trumpet and a violin, or a car engine, a piano or even the meow of a cat. The exhibition also offered a *Lissajous art gallery*. The visitor could change the subject and explore the different topics freely, without separated thematic compartments.

The scientific board members verified that argumental line and validated the topics, often adding critical remarks. For instance, we had conceived with the *Scale Lab* exhibit an interactive exploration of scales and tunings. This exhibit allowed the visitor to explore, measure, create and modify both individual sounds and musical scales. It covered the fundamentals of sound, perception of consonance and dissonance, and the creation of scales from single tones. However, one of our advisors pointed out that musical scales were not only based on acoustics; the notes in the Western scale had some functional relations that derive from tradition and cultural backgrounds, which lead to a combinatorial theory of scales. That led to some modifications in the *Scale Lab* exhibit, and to the creation of the exhibit *Note Compass*, which explored the modes of the diatonic (western) scale from that combinatorial perspective.

At the same time that we devised the main topics, we found it essential that the exhibition should contain true, actual music, and not be just about sound and acoustics. On this, again, the advisors helped out with some of their own prototypes and tools. With *Con Espressione!* the advisor and his research team had prepared an AI-assisted system to conduct a piano performance of classical music (Chopin, Beethoven, etc.), where the user can adjust tempo, loudness, and other parameters of the performance with hand gestures. We adapted that academic work to our exhibition by adding a user interface and a matching visualization. With this, we obtained both a true music exhibit and a display of the current research in the field. A similar example was *Tonnetz*, an exhibit that displayed a classical, graphical representation of notes and chords on a diagram that traces back to Euler ideas. One of our advisors was doing research on an interactive implementation of tonnetz ideas, with the ability to visualize live music or prerecorded musical pieces. We again adapted the user interface for an exhibition, making this another highlight exhibit with real music and real research. Other places where we included actual music were *Show Me Music*, which featured a variety of musical visualizations, and also in the form of short films and video clips in *La La Cinema*.

For our desire to have artworks, we had some success with the open call we launched on our web, but also by inviting artists we located. We obtained images, 3D sculptures, and other physical installations based on visualizations of harmonic, rhythmic, or acoustic patterns. Working with artists is definitely different than working with researchers, especially when it comes to the modification of their content or prototypes. Artists' contributions were installed according to their specifications, without significant adaptation from our side.

As an integral part of the development process, we designed and built the scenography of the exhibition. The MAINS showroom used to be an industrial machinery display, and it still had industrial power outlets, air pressure valves, tubes, and other obstacles over the walls and ceiling that we could not hide easily. We embraced that style and it was part of the *Lab* inspiration in *La La Lab*. However, we still needed to take over the space, to make the visitor feel immersed in an exhibition and not just a room with exhibits. Our cost-effective and impactful solution was to cover the lights with a colored film that made all the room illuminated in magenta glow that clearly framed the exhibition once one crossed the entrance curtains.

Another scenography issue was sound. With an exhibition about sound and music, it was essential to avoid interference between their sound, while avoiding having isolated spaces. We built some irregular walls using acoustic insulation fabric over a wooden frame, and we carefully arranged the exhibits with speakers and with headphones alternatively so the sound did not interfere much. Luckily (and a rare privilege), space was not a constraint, so we used almost 500 m^2 for our 20 exhibits. In later temporary installations, we managed to pack about 15 of those exhibits in less than 200 m^2.

(a)   (b)

Figure 8:   (a) and (b) *La La Lab — The Mathematics of Music* in Heidelberg, Germany, 2019. Image credit: Wanda Domínguez.

The exhibition (see Figures 8(a) and 8(b)) was open from May 2019 to March 2020, after an extension due to high affluence, but the 2020 COVID-19 pandemic halted all the touring venues that were originally planned. Some aspects could be further improved to achieve a science-museum-quality standard, such as casing/furniture for the exhibits, unity of style, and user interfaces across exhibits, etc. However, in terms of content and storyline we have explored the interlink between Mathematics and Music far beyond any public exhibition that we have found in existing museums.

For us, it was a challenging experience to find a balance between our own curation vision with the reality of what can be implemented, to balance playfulness with research depth exhibits, and to balance art with sound and mathematics. The exhibition is now starting to be duplicated, with locally produced exhibitions in Belgrade, Serbia in May 2021 and later in the year by a school in Bavaria, Germany.

I AM A.I. — A hybrid exhibition

I AM A.I. is our latest exhibition with 14 open-source exhibits explaining core concepts of artificial intelligence — and the mathematics behind it — through intuitive virtual and hands-on games. It was planned as a traveling exhibition starting in Heidelberg in May 2020. The exhibition turned into a digital exhibition[g] as part of a rapid response to the COVID-19 pandemic.

---

[g] https://www.i-am.ai/.

One of our strengths is to be able to adapt and respond to current needs when it comes to exhibition content, workshop content, and organization. We managed to turn part of the original *I AM A.I.* exhibition into an enhanced digital version. Within 1 month, we adapted a selection of the exhibits to be hosted digitally, developed a digital tour in both English and German, and managed to launch the digital online exhibition on June 10, 2020. The digital content comprises three educational games, a hands-on build your own exhibit at home, as well as a digital comic on AI. Just like in a physical exhibition, an interactive tour guide is also available to explain some of the content in more detail.

The name *I AM A.I.*, besides being a palindrome, appeals to a more emotional approach to an otherwise technical topic. The subject of the sentence, "I", is deliberately ambiguous. It can refer to the exhibition, in the sense that it performs a "human" task of explaining itself. It can be the visitor that reads the sentence, in the sense that we humans are intelligent, but our training data is the cultural baggage of humanity, which is by definition artificial. And it can also stand for how the "I" or "We" as a society are responsible for shaping artificial intelligence.

The exhibition, thus, tries to touch on several aspects of artificial intelligence: From one side, demystifying the technical aspects of AI, showing the inner workings of those systems in terms of simple mathematics, debunking the myth of the "magic" behind them. From the other side, it does not avoid more philosophical questions such as "what does it mean to be intelligent?", "Can an AI system have ethics?", or "What consequences can our society experience with the advent of more powerful AI systems?". All this mixed with some purely playful displays of systems using AI. We will now describe some of the exhibits and the rationale behind their role in the exhibition.

From the more technical and mathematical perspective, *Neural Numbers* provides a good entry example of what an AI system is. The task for this simple classifier is to identify your handwritten number from 0 to 9, which you write interactively into a box when exploring this exhibit online. Depending on how the network is trained, it will find it easier or harder to identify the written number. The ideas of training data, standardized input, and neural network architecture can be seen at a practical level.

A more detailed exploration of what exactly a neuron is, and how we train a neural network, is presented in *Simple Networks*, which presents a step-by-step tutorial on how neural networks are built from scratch. In that game, several neural networks are presented and the user has to train them (adjust the coefficients) by hand from some goal purpose, or some training data. Additionally, the network can be auto-trained by using the gradient descent algorithm, which is central in Neural Networks theory.

For this algorithm, we have the exhibit *Gradient Descent*, which requires the user to locate a treasure at the deepest point of a 1D seabed, by sending probes down to the seabed. The information returned from the probes provides the user knowledge of the value of the slope of the seabed at the point where the probe was sent down. This will allow the user to learn how following the slope downwards will eventually lead to the treasure. It also highlights how rugged seabeds search is much harder. In this way, the Gradient Descent algorithm, commonly used to train an AI, is introduced in an intuitive and entertaining way, but also highlighting issues these algorithms encounter in the real-life training of a machine learning model.

Other techniques of AI will also be present in the full, physical exhibition, for instance, *Reinforcement Learning*, which explains how a robot can find a route through a labyrinth by setting appropriate exploration and learning mechanisms. The physical exhibition was postponed to early 2022. Currently, it is traveling to different venues.

On the philosophical side, with *Build your own AI* we create a paper and cups system that can play the 2-player game of Nim. This shows that training an AI can in principle be done without a computer. Can just a few cups and papers have some intelligence on them? On the full exhibition, *Turing Game* develops the idea further by offering the user a full book of instructions on how to play the winning strategy in a more advanced version of Nim. This raises the question: does someone with the instructions book understand the game? Such a player is indistinguishable from another that has studied and understood the game by himself, or herself. This is a variant of the famous Turing test.

Other points for reflection in the exhibition are given in the graphic novel *We Need to Talk, AI*, a contribution by a researcher and illustrator from Germany, that gives a good introduction to the latest AI technologies, but also provides food for thought around moral topics in AI. Finally, the *Ethics of autonomous vehicles* explores the conundrum known as the Trolley Problem, in the context of automatic systems that can potentially decide on which is the least harmful outcome in life-or-death situations such as driving autonomously a car.

Despite having a digital exhibition, and a virtual guided tour, the experience for the visitor is never as rich as with a visit to a venue and a live tour with a mediator. Again, we followed the approach of going with the demand and ideas from visitors or web users and agreed to conduct online tours and online workshops. Based on our Science Spaces Workshops (see Figure 9(a)) [20], we established a 4-day long online workshop concept for students to learn about, experience and play with core concepts of AI. They included analog games (see Figure 9(b)), experiments, discussion games, interactive software (see Figure 9(c)), thought experiments, and analogies, mostly from our own exhibits from *I AM A.I.*, but also from other available resources with pedagogical value. These workshops were a success, and within several months we could offer more than 60 workshops in 17 countries, many of them held with partner schools of international Goethe Institutes. We built up a team of freelancers, and are now launching the whole workshop curriculum also under an open license, so others can conduct their own workshops too. Having experienced the potential of online workshops, we are now looking at hybrid exhibits whenever we design a new exhibition. This means there should ideally also be a version for online use (guided or for self-discovery). The online workshop work also brought us closer to formal education, where schools and educational institutes are now using our digital exhibits and planning to add these to local school curricula. *I AM A.I.* also was selected among 100 projects in innovation, education, and science by the accelerator program "Wirkung Hoch 100",[h] where it is accompanied to further scale and see how it contributes to a general system change in AI education and AI democratization.

---

[h] https://www.stifterverband.org/wirkunghoch100.

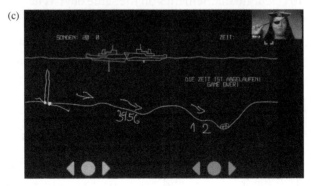

Figure 9: (a) Science Spaces workshop in Kigali, Rwanda, 2017. Image credit: Bianca Violet. (b) Workshop "AI explained" with students from the AI workshop in Tbilisi, Georgia, 2020. Image credit: Goethe-Institut Georgien. (c) Online workshop "AI explained" with students from New Delhi, India, 2020. Image credit: Bianca Violet.

Current ventures

Community-driven: The International Day of Mathematics

On March 14, 2020, the first-ever International Day of Mathematics[i] took place, celebrating the beauty and importance of mathematics in our everyday lives. This recurring event was proclaimed by UNESCO and initiated by the International Mathematical Union (IMU).

Every year, a new theme to flavor the celebration will be announced, sparking creativity and bringing light to connections between mathematics and all sorts of fields, concepts, and ideas. IMAGINARY was entrusted by IMU with designing and creating the International Day of Mathematics website, as well as the activities and resources to help people organize celebration events all over the world. Because of the COVID-19 pandemic, many planned events in countries all over the world have been downsized, postponed, or even canceled. As a consequence, a special Live Global Launch webpage accompanied all events worldwide on March 14, in 2020 and also in 2021.

For IMAGINARY, this project helped us to gain experience in a very large and decentralized, community-driven project. How can you motivate everybody to celebrate mathematics on the same day? What activities can you offer, that would work in all regions and for many age groups? We developed logos, poster sets, and activities such as a math scavenger hunt. Again, the format of competitions (or calls to participate) was successful, we generated a collaborative web video and received over 2,000 posters in our poster challenge.

Distilled format: The 10 Minute Museum of Climate Crisis

Exhibitions need a lot of space (and often also time to visit); they are heavy to transport, difficult to store, and expensive to produce. This inspired us to explore a new format for science communication, which we call a 10 Minute Museum[j]: a self-contained, free-standing, portable, physical station that offers the experience of a science museum exhibition in less

---

[i] https://www.idm314.org/.

[j] https://10mm.imaginary.org/.

than one square meter and in a brief amount of time. The project started in 2020 and comes along with an external evaluation study on the impact of the format and the knowledge transfer through observations (i.e., eye-tracking) and interviews with the museum's audience on site. The evaluation will be conducted by the German National Institute for Science Communication (NaWik) and will address the usability of the exhibition as well as its content. This is the first time where evaluation is part of our projects and we are looking forward to its results.

The 10 Minute Museums are a form of "pop-up" intervention of public spaces (libraries, shopping centers, tourist sites) that attract and surprise people, rewarding them with unexpected and enriching experiences during their daily activities. They entice through attractive construction and playful forms of interaction for a short but meaningful experience which can act as a starting point for a longer engagement (inviting to a museum exhibition, providing a link to a book, podcast, or mobile app, etc.).

Since the station is free-standing, it can offer access to content from all its sides, allowing for several people to directly interact with it at the same time. It's experienced by walking around it, as opposed to the static experience of a traditional "kiosk". The compact format with a variety of content forms and interaction possibilities dynamically arranged on its body invites exploration and discovery. It can combine traditional forms like text, images, sound, and video, and embedded display cases with touch screens, interactive lights, floor projection, or peepholes to watch stereographic images. It can provide mechanisms for visitor participation from a simple mail slot (for written or drawn feedback) to sound or video recording, and other richer interactions enabled by the visitor's mobile phone.

Our first 10 Minute Museum will focus on the science of the climate crisis with a strong emphasis on mathematical modeling. We aim to explain what climate models are, how they are built, and what they can tell us. We will also cover mathematical models that explain or predict the effects of climate change on plant and animal life, and human activities. Finally, we will present resources to dive deeper into the subject and ideas on how to take action from the political to the individual level.

Once it's completed it will tour several venues in Germany. We will share the experience of this new format through a free and open digital booklet that will explain it, give construction tips, guidelines, and insights. A 10 Minute Museum can be built on a variety of materials: custom-built wood casings, repurposed pieces of furniture or even cardboard boxes housing affordable electronics. We are excited by the potential of diverse groups (science communicators, NGOs, school classes, architects, designers, and artists) creating their own 10 Minute Museums and sharing their results with us.

Between mathematics and fantasy: Mathina

Mathina[k] is a 2.5-year collaboration Erasmus+ project launched in 2018 between IMAGINARY and four other partner organizations (Curvilinea from Italy, Atractor from Portugal, Experience Workshop from Finland and Bragi from Slovenia), to promote mathematical thinking to children and students of all ages. The project develops a web platform with fictional stories where the protagonists — Mathina, a 15-year-old girl, and her brother Leo — travel around a fantasy land and find mathematical challenges and puzzles that the reader can interact with, reflect on, and learn from, in the form of interactive apps.

Mathina is a somehow experimental format, that at the moment of writing we are still testing with focus groups of students. Mathina can be used by parents to play with their kids, by teachers as a didactical tool, or directly by young learners eager to discover more math. The math content on each story is deliberately more advanced than what would correspond in the school for that age. This aims to provide the reader with an intuitive, early exposure to those topics, in line with non-formal learning strategies. In Mathina's world all learners are adventurers, who can actively participate in the discoveries and autonomously reflect on their own learning journey.

A completely new feature for us in this project is hand-drawn illustrations, which give the whole story soul and a very warm touch (see Figures 10(a) and 10(b)). We are very curious to see how this form of interactive storytelling combines with mathematical knowledge transfer [17].

---

[k] https://www.mathina.eu/.

(a)

(b)

Figure 10: (a) Overview map of the adventure land of Mathina. (b) Screenshot from within one of the stories taking place in the land of birds of fire.

Other experiments and tools

We love to experiment and test projects and formats. In this spirit, we constantly develop small apps, widgets, or gadgets, which we use online or offline. Examples are web applications for a *New York Times* article for Valentine's Day [16], online games such as Space Rescue[l] or a mini-website for e-day.[m] We also prepared stickers of pentagonal tilings or temporarily tattooed workshop visitors with self-created algebraic surface tattoos.

Larger projects include a survey among 30 mathematics museums on the question "What is the best math exhibit?" with the National Museum

---

[l] https://imaginary.github.io/space-rescue/.

[m] http://eday.imaginary.org/.

of Mathematics, or the open calendar map Pop Math, with the European Mathematics Society.

New projects are a dynamic infrastructure for galleries and museums to quickly curate exhibitions (project museum4punkt0) and a hands-on station with Lego bricks and projections on the future of mobility for the Futurium museum in Berlin.

A Never-Ending Journey

We believe that the word "journey" fits well to IMAGINARY, its history and its current projects, and probably also to the future. We are "on our way" and constantly adapt how we communicate mathematics. And this way, we hopefully also continue shaping mathematics communication and engagement, and how the world perceives and — more importantly — interacts with mathematics in the long run. The demand is there as we can see in the science centers and many non-formal learning opportunities that emerge and are welcomed and frequented by the many visitors.

In accordance with this increasing demand for resources to explore and understand science and mathematics in non-formal environments that parallel schools, and from a public increasingly interested in a scientific culture, exhibitions will be part of those resources, but many other formats will also be demanded.

IMAGINARY hopes to keep in balance with that demand. That will most probably require improving the build quality of exhibits, the flexibility of the setups, and establishing a distribution flow in which our exhibitions can be re-used easily while allowing us to innovate and produce exhibitions and resources at a steady rate. At the same time, we look for arrangements that allow us to keep our background academic perspective, where we can embrace exhibition and communication projects by their intrinsic value. In this direction, we are exploring agencies and partnerships with education and research institutions that can support and shape the future of IMAGINARY with us.

From the very first exhibition until now, we are open for collaboration and also offer everything we do under open licenses. We love to experiment with new formats and new technologies, we love to communicate research mathematics and we try our best to be fast and pragmatic while listening to our public and the community. We are very grateful to all our supporters and collaborators, who accompany us on this great journey.

Acknowledgments

We thank our supporters: the Klaus Tschira Foundation for the open-source platform, the snapshots project, the La La Lab exhibition, the International Day of Mathematics, and the Climate Crisis museum (2011–2021), the BMBF for the original IMAGINARY exhibition, and the traveling exhibitions in Turkey and Israel (2007–2013), the Leibniz Association for the foundation of our own nonprofit organization (2016–2018), UNESCO for projects in more than 7 African countries (2014–2018), and the Carl-Zeiss Foundation for the *I AM A.I.* physical and digital exhibition (2020–2022).

IMAGINARY only lives, because of all collaborators, co-workers, freelancers, volunteers, contributors, and local organizers of exhibitions and projects. Most of the projects mentioned above are only possible because of you, your ideas, and your work. Thank you!

References

[1] Aldrighi, P. C. (2018). *IMAGINARY — Un viaje por la matemática* [*IMAGINARY Uruguay: A Mathematical Journey*]. IMAGINARY, https://www.imaginary.org/film/release-of-the-documentary-film-imaginary-uruguay-a-mathematical-journey. [Accessed on June 17, 2021].

[2] EPSRC Centre for Doctoral Training (CDT) in the Mathematics of Planet Earth (MPE). (2017). *MPE Catalogue*. London. http://mpecdt.org/wp-content/uploads/2017/10/IC-MPE-Exhibition-2017-brochure-ARTWORK.pdf. [Accessed on June 17, 2021].

[3] Greuel, G.-M. and Matt, A. D. (eds.) (2014). *IMAGINARY Catalogue*. Travelling Exhibition Catalogue. IMAGINARY, Mathematisches Forschungsinstitut, https://www.imaginary.org/sites/default/files/imaginary-catalogue-edition-2014_0.pdf. [Accessed on June 17, 2021].

[4] Grünberg, D. and Matt, A. D. (2015). "Using surfer to investigate algebraic surfaces". *The Mathematics Teacher*, 109(3), 221–226.

[5] Hartkopf, A. and Matt, A. D. (2013). "Surfer in math art, education and science communication". *Bridges Proceedings*, 271–278.

[6] IMAGINARY. (2019). *La La Lab — exhibition booklet*. https://www. imaginary.org/sites/default/files/20190911-lala-booklet-v0.4-web-text.pdf. [Accessed on June 17, 2021].

[7] La Real Sociedad Matemática Española (ed.) (2011). *IMAGINARY — Una mirada matemática*. GRÁFICAS PAPIRO. *IMAGINARY*, https:// www.imaginary.org/sites/default/files/rsme-imaginary_catalogue_es.pdf. [Accessed on June 17, 2021].

[8] Matt, A. D. (2017). "IMAGINARY — a how-to guide for math exhibitions". *Notices of the American Mathematical Society*, 64(04), 368–373.

[9] Matt, A. D. (2012). "IMAGINARY and the Open Source Math Exhibition Platform". *Raising Public Awareness of Mathematics* (pp. 167–185). Berlin, Heidelberg: Springer-Verlag.

[10] Matt, A. D., *et al.* (2013). "How to make an IMAGINARY exhibition". *Bridges Proceedings*, 389–396. https://www.imaginary.org/background-material/ how-to-make-an-imaginary-exhibition-bridges-2013-paper. [Accessed on June 17, 2021].

[11] Matt, A. D., *et al.* (2018). "Modern mathematics communications — An Asian focus". *Asia Pacific Mathematics Newsletter*, 8(1). IMAGINARY. https://www.imaginary.org/background-material/modern-mathematics-communications-an-asian-focus.

[12] Matt, A. D. and Grünberg, D. (2014). "IMAGINARY-Entdeckerbox für Schulen". *Beiträge zur 48. Jahrestagung der Gesellschaft für Didaktik der Mathematik*, WTM Verlag für wissenschaftliche Texte und Medien Münster.

[13] Matt, A. D. and Violet, B. (2016). "Collaborative Mathematics Communication–Experiences and Examples". *ICMI-13 post-congress monograph of Topic Study Group: Popularization of Mathematics*. IMAGINARY. 13th International Congress on Mathematical Education, https://www.imaginary.org/ background-material/collaborative-mathematics-communication-experiences-and-examples. [Accessed on June 17, 2021].

[14] Pöppe, C. (2008). "Mathematik-Kunst-Wettbewerb: die Ergebnisse". *Spektrum der Wissenschaft*, 8(8), 2008 94–97.

[15] Ramos, D. and Violet, B. (2020). "La la lab — The mathematics of music". *Bridges Proceedings*. IMAGINARY. https://imaginary.org/background-material/la-la-lab-the-mathematics-of-music. [Accessed on June 17, 2021].

[16] Roberts, S. (2019). "The Perfect Valentine? A Math Formula". *New York Times*, 14 February 2019. https://www.nytimes.com/2019/02/14/science/ math-algorithm-valentine.html. [Accessed on June 17, 2021].

[17] Somlyódy, N. and Fenyvesi, K. (eds.) (2020). *Mathina Program Booklet — Mathina's World*. Mathina, https://mathina.eu/documents/Booklet_Mathina. pdf?v=1584032400. [Accessed on June 17, 2021].

[18] UNESCO. (2013). "MPE2013 moves into Mathematics of Planet Earth". http://mpe.dimacs.rutgers.edu/wp-content/uploads/2013/12/press_release_ EN_Dec11.pdf. [Accessed on June 17, 2021].

[19] Violet, B., *et al.* (2017). "Math creations — A math-art competition". *Bridges Proceedings*. IMAGINARY. https://imaginary.org/background-material/ math-creations-a-math-art-competition. [Accessed on June 17, 2021].

[20] Violet, B. and Damrau, M. (2018). "Science spaces: An open workshop concept to create science exhibits". *Bridges Proceedings*. IMAGINARY. https:// imaginary.org/background-material/science-spaces-an-open-workshop-concept-to-create-science-exhibits. [Accessed on June 17, 2021].

[21] Violet, B. and Matt, A. D. (2016). "Mathematikon: A mathematical shopping center". *Bridges Proceedings*. IMAGINARY. https://imaginary. org/background-material/mathematikon-a-mathematical-shopping-center. [Accessed on June 17, 2021].

https://doi.org/10.1142/9789811253072_0010

Chapter 9

Science Communication and Outreach Events During the *Illustrating Mathematics* Semester Program at the Institute for Computational and Experimental Research in Mathematics (ICERM)

Martin Skrodzki

ICERM, Brown University, Providence, RI, USA; RIKEN iTHEMS,
Wako, Saitama, Japan; CGV, TU Delft, Delft, the Netherlands
mail@ms-math-computer.science

Good illustrations of mathematical content enhance the science communication of mathematics as well as the scientific process itself. However, the creation of these illustrations, the rendering of videos, the process of *mathematical making*,[a] is still underdeveloped and undervalued by many professional mathematicians.[b] The semester program *Illustrating Mathematics* aimed at closing some of the gaps between the richness of mathematical content and the availability of corresponding illustrations by creating new and novel ways to show and present mathematics.

This article reports on the program's related science communication and public outreach activities. To present the frame in which the activities took place, the first two sections briefly introduce the institutional setup and the scientific aspects of the semester program. The following sections each present one outreach activity in detail.

ICERM

In 2010, the five mathematicians Jill Pipher, Jeffrey Brock, Jan Hasthaven, Jeffrey Hoffstein, and Bjorn Sandstede founded the Institute for Computational and Experimental Research in Mathematics (ICERM).[c] This was possible through financial support by Brown University, the National

---

[a] Creation of illustrative content that is outside the traditional scope of visualization. It can comprise of, but is not limited to, literal objects, such as sculptures, paintings, or fabrics, but also digital images, software, or even performance arts.

[b] This is the underlying problem addressed in the *Mathematical Maker's Manifesto*. See: Farris, F. (2020). "Where does 'mathematical making' fit in our community?" *Notices of the American Mathematical Society*, 67(5), 614–615, signed by 30 participants of the *Illustrating Mathematics* program.

[c] Find more information about the institute on its website: https://icerm.brown.edu/.

Science Foundation (NSF), and the Division of Mathematical Sciences. The goal of ICERM is to promote computational methods and experiments in mathematical research. It clearly says in the mission statement of ICERM:

> "The mission of the Institute for Computational and Experimental Research in Mathematics (ICERM) is to support and broaden the relationship between mathematics and computation: specifically, to expand the use of computational and experimental methods in mathematics, support theoretical advances related to computation, and address problems posed by the existence and use of the computer through mathematical tools, research and innovation."[d]

ICERM executes this mission by hosting up to two semester programs (each with 3–4 affiliated workshops), several week-long topical workshops, and a summer undergraduate research program every year. Aside from the board of directors, ICERM has a permanent staff of eleven people. They take care of, for example, finance, available computer systems and software, as well as marketing and ICERM's internet presence. Except for at least one public lecture per semester, ICERM does not generally engage in outreach activities. Every summer the *Girls Get Math* event provides education for girls, but this event is not open to the public nor directed at a general audience. In 2019, ICERM was in its second 5-year NSF grant. These grants do not cover outreach or science communication activities. Still, the *Illustrating Mathematics* semester program was a welcome opportunity to engage in such events.

The *Illustrating Mathematics* Program

From September to December 2019, the semester program *Illustrating Mathematics*[e] took place at ICERM. This specific program brought together mathematicians, makers, and artists to find new forms of illustrations in and for mathematics.[f] Counting all workshop attendees and members of the whole semester program, 257 people participated in the various program

---

[d] Mission statement taken from https://icerm.brown.edu/about/.

[e] Find a description of the program and its various activities here: https://icerm.brown.edu/programs/sp-f19/. Frank Farris describes a previous workshop with the same title in an article in 2017, "With new technology, mathematicians turn numbers into art", which is available at https://scitechconnect.elsevier.com/new-technology-mathematicians-turn-numbers-into-art/.

[f] This website contains a list of projects from the program: https://im.icerm.brown.edu/.

activities. Through a generous grant from the Alfred P. Sloan Foundation, this semester program was able to reach out into the local community of students as well as artists. Thus, the program got the opportunity to present its contents to a broader audience and not only to the visiting mathematicians. The program website states:

"Objects created for mathematical visualization are beautiful and attractive in their own right. '*Illustrating Mathematics*' brings together artists, makers, and mathematicians seeking to harness the creativity of mathematical illustrations to further the public's understanding of mathematical research. ICERM invites you to engage with these talented artists-in-residence and explore their work. Visualize mathematics through displays of art made using 3D printing, laser cutting, CNC routing, virtual reality, textiles, carving, painting, video, and more!"[g]

There were several events revolving around public outreach and mathematics communication during the program. They included an on-site and an off-site mathematical art exhibition, an ICERM open house during a STEM-themed citywide *Waterfire*[h] event, and five Math + Art panels featuring artists and mathematicians answering questions from the audience about how art and math relate. During *Waterfire*, and again later in the semester, the public were invited to participate in building two large mathematical sculptures in ICERM's modern and open space. Diana Davis, one participant of the *Illustrating Mathematics* program, edited a book, which collected the projects that the various participants created throughout the semester. The remainder of this article provides a detailed discussion of these events with a special focus on their science communication aspects.

Math + Art Exhibitions

Two exhibitions enabled the public to engage with art inspired by mathematics. Topics ranged from discrete and differential geometry over

---

[g] Public Outreach statement taken from https://icerm.brown.edu/programs/sp-f19/#publicoutreach.

[h] *Waterfire* is an event in Providence, RI, that takes place multiple times annually. It has attracted millions of visitors to come to downtown Providence, to see the burning fires in braziers on the water, and to interact with the local community. For more information, see https://waterfire.org/.

topology to combinatorics. The mathematical content was usually encoded in the art object and was not immediately accessible to the audience. Furthermore, the artist might reduce the underlying mathematical structure for either practical[i] or artistic[j] purposes. The exposition of the art did not include explanatory texts but only the title of the artwork and the name of the artist. The organizers chose this presentation style to create the feeling of a typical art exhibit. The mathematical background of the presented artworks was quite complex, generally not possible to understand from the artwork itself, and mostly required more than a page of introduction. Therefore, to prevent the conversion of the event from an art exhibit into a small-scale mathematical museum, the organizers only showed the reduced information for each object.

Both exhibits ran parallel to the semester program. While one exhibition was located at ICERM from September to November 2019, the institute coordinated with the Brown Arts Initiative to host a second, more formal exhibit at the *Granoff Center*, on the campus of Brown University (see Figure 1). The center serves as a place for teaching activities in the arts and as a hub for collaboration among the arts, sciences, and humanities.[k] Over the course of about 4 weeks, mostly students, faculty, and staff passed through the exhibition where the different art pieces confronted them with the underlying mathematics.

Two receptions at the *Granoff Center* enabled the interested public to engage with the artwork and the present artists. Despite open calls, the audience comprised almost completely of art students from the *Granoff Center*, artists from the local community, and members of the *Illustrating Mathematics* program. Smaller groups formed in front

---

[i] As an example for practical limitations, in the work *2-adic Solenoid* by Dina Buric, it is impossible to print the infinitude of small twisted fibers that form the solenoid. The gallery of accepted artworks lists the solenoid here: http://gallery.bridgesmathart.org/exhibitions/2019-icerm-illustrating-mathematics/buricd.

[j] In her work *78 paths to decompose a sphere*, the artist Silviana Amethyst Brake cuts the ongoing paths around a sphere to focus on the center. Thereby, the actual mathematical paths outside the print are lost to the observer. Find the work here: http://gallery.bridgesmathart.org/exhibitions/2019-icerm-illustrating-mathematics/danielleamethyst.

[k] Find more information on the *Granoff Center*, its mission, and its activities here: https://arts.brown.edu/granoff-center.

Figure 1:   Part of the Exhibition at the Granoff Center. Credit: Mark. J. Stock.

of the artworks and discussed the underlying mathematics, sometimes with the artists themselves, sometimes with other mathematicians present, and sometimes among themselves as laypersons. The exhibition at ICERM had similar effects: workshop visitors and visitors to the Math + Art Panels (see below for a separate discussion of these events) attended receptions taking place in the exhibition space (see Figure 2). Many visitors then wandered around and explored the exhibit, which gave the artists — if present — an opportunity to explain their artwork and subsequently also the underlying mathematics. The exhibition catalogue[1] printed at the end of the semester contained texts by the artists describing their works, and therefore presents a helpful tool to understand what mathematical theory inspired the artwork and how the artist chose to present this theory.

The organizers of the exhibits did not install any success measures or feedback mechanisms. In terms of the challenges regarding the setup of the exhibitions, they reported that these were only possible because of

---

[1] Bachman, D., Schleimer, S., and Segerman, H. (eds.) (2019). *Illustrating Mathematics Art Exhibit Book — 2019*. Tessellations Publishing, Pheonix: ICERM.

Figure 2: Part of the exhibition at ICERM. Credit: Laura Taalman.

the online *Mathematical Art Galleries*[m] infrastructure that many of the participating artists were familiar with. Setting up such infrastructure would have overstrained the locally available resources and thus would have rendered the entire project infeasible. The synergy of using an established curating system and not having to set up a new way of collecting submissions thus saved a tremendous amount of time and energy and therefore enabled the presentation of the mathematical art in the first place.[n]

---

[m] The galleries started as a place to organize the annual art exhibit at the Bridges Conference (https://bridgesmathart.org/), but as of now also includes the exhibits at the Joint Mathematics Meetings (JMM, https://jointmathematicsmeetings.org/jmm) and two spin-offs of the Bridges Conference: a short film festival and a fashion show. Find the gallery devoted to the *Illustrating Mathematics* program here: http://gallery.bridgesmathart.org/exhibitions/2019-icerm-illustrating-mathematics.

[n] The background information on ICERM, the *Illustrating Mathematics* program, and the two exhibitions come from the presented sources and from a personal conversation with J. Elisenda Grigsby, Deputy Director of ICERM.

Big Bang Science Fair During *Waterfire*

The *Waterfire*° open house event enabled the public to engage with different mathematical (art) material. The activity at ICERM was part of the *Big Bang Science Fair*, which is an event that celebrates the intersections between science and the arts. The target group was both kids and adults who enjoy hands-on activities to discover science. At ICERM, the semester participants ran the event with their material and their prepared activities. Several tables represented different mathematical topics: *two-dimensional geometries, three-dimensional geometries, mathematics in motion*, and *computational textiles*.

Each of the long tables held a multitude of mathematical illustrations and objects. The visitors were free to interact with all of these objects, i.e., lift them up, rearrange them, or use them to build objects that are more complex than their basic parts. Mathematicians behind the tables answered questions and explained the underlying mathematics of the presented exhibits. Children in particular showed great interest in the mostly colorful objects and in the possibilities of what to build from them (Figure 3).

Over the course of 2 hours, visitors interacted with the various objects and discussed with the available mathematicians. Aside from the four tables, the visitors could participate in a large-scale building activity to construct an illuminated *stellated dodecahedron* (Figure 4), organized by semester program participant Glen Whitney. These hands-on activities serve a twofold pedagogical purpose. First, the opportunity to build something themselves sparks the interest of children and adults alike. Second, it is easier to illustrate mathematical facts on some physical objects than just describing the mathematics. For instance, the visitors can easily count the number of vertices, edges, and faces of the large dodecahedral sculpture. From this information, it is possible to illustrate the Euler characteristic and thus give a simple example for a deep mathematical result. Furthermore, the illumination of the sculpture indicated the different symmetries encoded in the geometrical object.

° See footnote h for a short description of the event.

Figure 3: Edmund Harris explains how to form complex objects from basic Curvahedra tiles at the open house event during the Big Bang Science Fair. Credit: Ruth Crane.

Figure 4: Left: Construction of the stellated dodecahedron at the open house event during the Big Bang Science Fair. Credit: Ruth Crane. Right: The finished and illuminated sculpture. Credit: Saul Schleimer.

A main goal of the participation of ICERM in the *Waterfire* event was raising awareness about the existence of the institute among the local residents of Providence and — on a larger scale — Rhode Island. While the local community knows Brown University as a whole, its different institutes are generally unknown. An event like the *Waterfire* is a great opportunity for smaller university entities — like ICERM — to make themselves known and to promote their research results. Toward this end, the event was a tremendous success for the institute. Because the semester program participants presented the materials they had produced so far or brought from home, preparation times were very short. The organizers of the *Big Bang Science Fair* advertised the event. In total, 300 free tickets to visit the ICERM open house were available and within hours, the distributors handed out all tickets. The public demand for more tickets would have been there, given the tens of thousands of participants in the surrounding event, but the location simply could not host more attendees. However, even after the event ended, the mathematical sculpture continued to shine from the eleventh floor onto the visitors of the *Waterfire* event on the street below, thereby raising attention for ICERM.

Building Activity *Math's Bubbling (not!) Over*

Toward the end of the program, Glen Whitney organized another large-scale building activity open to the public in the conference room at ICERM titled *Math's Bubbling (not!) Over*. Throughout the course of a day, participants built a model of the *Weaire-Phelan (W-P) foam*[p] from wooden dowels, custom plastic connectors, and acrylic cutouts for the faces of the foam cells. The organizer of the event writes on the goals of his activity:

> "*Math's Bubbling (not!) Over* was originally planned to be built and displayed as a part of a public presentation about mathematics. In that context, the goal of the construction was to illustrate how discovering

---

[p] The *W-P foam* is a hypothesized structure for an accumulation of a large (technically, infinite) number of bubbles of equal volume. Theoretically, such an accumulation would divide all of space into cells of equal volume with the minimum amount of surface area per cell. However, it remains unknown whether the *W-P foam* actually achieves the minimum possible surface area per cell.

a mathematical pattern or phenomenon is rarely the end of a story, but typically leads to further layers of understanding and appreciation and new questions, in a potentially endless cyclic[q] process. (I wanted to contrast this with the notion that a mathematician encounters a problem and then finds 'the solution' and then is done.)

Thus, confronted with a collection of line segments of specific lengths, the audience could discover that they could fit together to make certain pentagons and hexagons. Now equipped with those polygons, the audience could discover they created certain polyhedra, and then in turn that those polyhedra would fit together to tesselate all of space[r], see Figure 5. Finally, the audience would be in a position to understand and appreciate the unsolved problem of whether this division of space is optimal in Kelvin's sense.[s]

However, the public presentation was not able to take place for personal reasons. Hence, the installation as it actually occurred became an opportunity for members of the ICERM community to converse in the context of a shared project, to experiment with construction techniques, and to observe the *W-P foam* at an unusual scale and from unusual viewpoints (e.g., from the interior of a cell). Based on feedback from participants, it seems that the installation succeeded in these altered goals."

Aside from the goals of Glen Whitney, the building activity provided another means for ICERM to interact with the community and gain additional visibility. In terms of mathematical communication, it was an opportunity for members from the semester program to interact — given their own expertise — on a lesser-known subject of mathematics. Thereby, the building activity established communication between different subdisciplines of mathematics.

---

[q] While Glenn Whitney spoke about a "cyclic" process, it is safe to assume that he meant to describe it as a spiraling process that revisits certain questions while still going forward, unlike a cyclic repetition that always comes back to the exact same starting point.

[r] That is to say that the entire space was covered without neither holes nor any overlap between the covering objects.

[s] As stated in footnote p, it is unknown whether the *W-P foam* achieves the minimum possible surface area per cell. Sir William Thomson (later Lord Kelvin) investigated this minimization problem in 1887.

Figure 5: The fully constructed *W-P foam* model. Credit: Edmund Harris.

Math + Art Panels

The origins of this activity originated through efforts of Richard E. Schwartz (mathematician at Brown University), Masha Ryskin (associate professor at the Rhode Island School of Design, RISD), and Allison Pashke (Rhode Island based artist). They aimed at involving RISD and the local art community in the semester program. ICERM and RISD are within walking distance but did not have any strong ties before this project. Schwartz frequently gave guest lectures on geometry in Ryskin's classes at RISD. Together with Allison Pashke, they came up with the idea to host panels at ICERM.

Throughout the *Illustrating Mathematics* semester, five panels were hosted, each of them scheduled for the duration of 1 hour. The members of a single panel consisted of two organizers and four participating artists and mathematicians. After a short introduction by the organizers, every participant had 5 minutes for a short presentation of her or his work. Discussions between the audience and the panelists followed these short talks and the panels usually went about 15 minutes over their 1-hour time limit.

When preparing the panels, the organizers aimed for a variety of speakers. They abandoned a first idea to fix a topic for each panel and to invite corresponding panelists because it turned out to be too logistically complicated. Similarly, the organizers abandoned the idea of having exactly two artists and two mathematicians speaking in every panel. This is also because for most participants in the semester program, the organizers could not find good pairings of artists and mathematicians. The remaining goal was to provide a variety of speakers in each panel and to have at least one working mathematician speaking in each panel. The invitation process of speakers for the panel revealed a culture clash between the mathematics and the arts/design community. The two communities handled both the selection process and — even more important — the question of whether or not to pay the speakers differently. The second aspect of this conflict had to be resolved by finding funds to pay the speaking artists, which was possible by the grant of the Alfred P. Sloan Foundation. In regards to the first aspect, Masha Ryskin and Allison Pashke compiled a list of possible arts and design panelists, while Richard E. Schwartz compiled a list of possible mathematicians for the panel, thereby taking the varying selection criteria into account for both sets of speakers.

Early on during the planning, the organizers decided to overlap several of the panels with the workshops at ICERM. This had several benefits. On the logistics side, each workshop had a starting reception in the common area of ICERM. By scheduling the events right after these receptions, the interested public could engage the mathematicians and artists attending the workshop. Furthermore, the visitors had an opportunity to see the arts exhibition at ICERM (see above for a discussion of the exhibition). Richard E. Schwartz remarked that overlapping the panels with the workshop reception also gave a "classy air"[t] to the panels while the workshop participants provided a "fresh audience" to every panel. This contributed greatly to the discussions after the short presentations, driven by the audience. These discussions were different in every panel, ranging from very work-specific ("What problems did you encounter when cutting these wood pieces with a laser cutter?") to rather

---

[t] Quotes from a personal conversation with Prof. Richard Evan Schwartz, Brown University.

Figure 6: David Bachman introduces the panelists of the second Math + Art panel. Credit: Ruth Crane.

broad and general questions ("When do you consider a piece 'done'?"). See Figure 6 for a picture of one of the Math + Art panels.

Aside from the workshop participants, the organizers aimed their panels at interested students and faculty from both Brown University and RISD. It turned out to be very beneficial to have more than one panel as the event quickly established itself in the local art community and about a dozen local artists came to the last panel. Each panel had an audience of about one hundred people, which surpassed the expectations of the organizers, who initially worried that not enough participants would come.

In summary, the organizers acknowledge that the event would be hard to repeat without ICERM as a hosting institution. The logistics, support in handling speakers and invitations, rooms, catering, and paying of the artists by ICERM (with help from the Alfred P. Sloan Foundation) proved invaluable and crucial for the whole activity. In hindsight, the panels themselves went too quickly and there was not enough time to answer the questions in sufficient depth. An immediate success of the panel series was the start of several cooperative projects — both artistic and mathematical — between members of the audience and panelists. Furthermore, many students came to the panels and became interested in their contents, which Richard E. Schwartz sees as another big success. While he does not necessarily want to run another panel series, he surely wants to continue with other "lively math events for the public".[u]

Of course it is not surprising that the co-organizer of the panels considers his event series a success. However, the sheer fact that the organizational efforts were too large to continue the series speaks for itself. Obviously, the gains do not compensate the efforts put into the organization. The questions show that the mathematical content was at most shallow, if present at all. Time did not permit for detailed discussions and while all of the images had beautiful artwork inspired by mathematics, it has doubtful that any actually mathematics were communicated during these events. Furthermore, while students and members of the local art community were present, these were — for the most part — mathematics enthusiasts, which meant that the panelists were preaching to the choir. In this sense, the event did fail to reach out in the broader local community. Further evaluation and, for instance, a questionnaire handed out to the participants would shed some light on these aspects. Sadly, this opportunity is gone, but should be picked up in any similar event.

The *Illustrating Mathematics* Book

A final outreach project that originated in the *Illustrating Mathematics* program is an eponymous book, edited by Diana Davis.[v] Unlike an exhibition book that solely presents the objects and the artists, this book focuses on the medium used for the work as well as on the process of making the object. Therefore, it does not only include images of the final creations, but also stories from the process, like things that went wrong or got changed along the way. Guiding points from the authors elaborate on the medium, the mathematics on display, why the person chose this content, why they chose

---

[u] Quotes and background information on the Math + Art panels originate in a personal conversation with Prof. Richard Evan Schwartz, Brown University.

[v] Davis, D. (ed.) (2020). *Illustrating Mathematics*. Providence, Rhode Island: American Mathematical Society.

this medium to illustrate the content, what the creator learned from making the object, and what the observer can learn from the result.

The book accordingly has eight sections, each presenting works with a different medium. They consist of *Drawing, Paper & Fiber Arts, Laser Cutting, Graphics, Video & Virtual Reality, 3D Printing*, and *Mechanical Constructions & Other Materials*. The final section, *Multiple Ways to Illustrate the Same Thing*, explores how different media can capture different aspects of a given mathematical concept. The contributions highlight how different illustrations can complement and complete each other in the task of conveying its underlying mathematics. In the introduction, Diana Davis writes that her book does three things. It "showcases the great variety of materials for illustrating mathematics, gives voice to people's stories about illustrating their mathematics, so that we can learn from their experience, and shows the variety of ways that different people use the same materials in very different ways."[w]

Figure 7 shows a representative page with the illustration work of the author of this article.[x] The illustrated mathematical content is the process

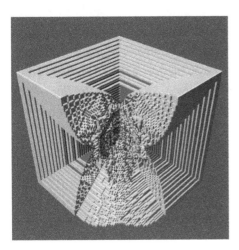

Figure 7:   A page from the *Illustrating Mathematics* book. It illustrates the process of three-dimensional chip firing. Image Credit: Caroline J. Klivans/Pedro F. Felzenszwalb/Martin Skrodzki. Credit: American Mathematical Society.

---

[w] *Illustrating Mathematics*, p. 3.

[x] *Ibidem*, pp. 62–63.

of *chip firing* on an undirected graph network, sometimes also referred to as a *sandpile*. Starting with an initial stack of chips at a given starting position in the network, the process distributes these chips along the connections of the network. In each step, if possible, each network node distributes one chip to each of its connected neighbors. For a large enough graph, eventually, the process ends and forms intricate structures like the one shown in the left part of Figure 7. The book of Diana Davis collects 70 such illustration projects, thereby serving not only as a chronicle of the *Illustrating Mathematics* program, but also as a source of inspiration for both further illustrations and upcoming illustration programs.

Final Thoughts and Outlook

While the activities presented in this article are the main events that took place during the *Illustrating Mathematics* semester at ICERM, there are more science communication projects to come. Mikael Johanson gave a talk on how to create Twitter Bots that present mathematical content automatically. One result is the bot of Dina Buric, which now posts daily fractal images based on the Tribonacci graph.[y] Evelyn J. Lamb, a freelance math and science writer, gave a talk on mathematical writing for the public. The articles in her blog are great examples.[z] Steve Trettel and Sebastian Bozlee explored how to interact with mathematics in virtual reality.[aa] These and more science communication projects originating from the *Illustrating Mathematics* semester at ICERM will be available in the coming months and years.

Acknowledgments

This material is based upon work supported by the National Science Foundation under Grant No. DMS-1439786 and the Alfred P. Sloan

---

[y] Find the bot of Dina Buric at https://twitter.com/GraphTrib/. Other mathematical twitter bots include tiling bot (https://twitter.com/tilingbot) by Roice Nelson and a bot on symmetric curves by Mikael Johanson (https://twitter.com/symmetric_curve), inspired by Frank Farris.

[z] http://www.evelynjlamb.com/.

[aa] Sebastian's work is featured in several videos on YouTube, e.g., https://www.youtube.com/watch?v=N_bQ94bGHE0, while Steve's work can be interactively explored on his webpage http://www.stevejtrettel.site/.

Foundation award G-2019-11406 while the author was in residence at the Institute for Computational and Experimental Research in Mathematics in Providence, RI, during the *Illustrating Mathematics* program.

The author would like to thank Ruth Crane, Eli Grigsby, Richard Evan Schwartz, and Glen Whitney for their contributions and many helpful discussions about this article. Further thanks goes to Diana Davis, editor of the *Illustrating Mathematics* book and to the organizers of the *Illustrating Mathematics* semester, without whom none of these activities would have taken place: David Bachman, Kelly Delp, David Dumas, Saul Schleimer, Richard Schwartz, Henry Segerman, Katherine Stange, and Laura Taalman.

https://doi.org/10.1142/9789811253072_0011

Chapter 10

Visualization and Social Media as Tools for Mathematics Communication: An Account of the Project "Sketchnotes of Science"

Constanza Rojas-Molina

Lecturer, CY Cergy Paris Université, France
hello@crojasmolina.com

Introduction

Drawing is part of our lives from early childhood and is transversal to age, gender, social, and educational background. Drawing lines and shapes lays the ground to learning to write, and it comes in a natural way to help us process the world around us and express ourselves where words are lacking. This is what makes drawing so powerful and appealing in all ages of development. In this note, we present the project "Sketchnotes of Science", carried out by the author, where drawing and mathematics are at the center of the activities. Here, we take drawing and use it to invite the audience to reflect with us on mathematics, full of abstract and complex logical constructions, and the people who construct them, the mathematicians.

The project "Sketchnotes of Science" consists of two drawing challenges in social media (Twitter and Instagram) carried out in collaboration with other communicators. The term "sketchnotes", coined by designer Mike Rohde,[a] denotes a visual approach to notetaking. This is a combination of doodling, or sketching, and the traditional act of taking handwritten notes during a lecture, a presentation, or any situation where a person is receiving and processing information. This approach naturally leads to increased focus, better retention of the information, and the visual aspect makes it

---

[a] https://rohdesign.com/.

appealing to other people, which motivates the sharing of information.[b] In the words of Mike Rohde, sketchnotes is about "Ideas, not Art". This means that it is not necessary to be a talented artist to do sketchnotes, but simply to be able to convey ideas through a visual language that can be learnt and refined via practice. Sketchnoting in the form of drawing challenges is turned into a collective activity (social media) where everyone interested can challenge themselves to participate and publicly share their output. The format of a drawing challenge takes inspiration from the drawing challenge "Inktober" by illustrator Jake Parker.[c] This challenge, well-known in the illustration community, takes place each year in October, and its goal is to create drawing habits among illustrators. The project "Sketchnotes of Science" combines therefore sketchnotes and the drawing challenge format to make the audience interested in mathematical concepts and in the people who do mathematics, with an emphasis on the ideas, not on the art. The sketchnotes are promoted on social media and on the blog *The RAGE of the Blackboard*, written and illustrated by the author.

In the following we detail the two subprojects that compose "Sketchnotes of Science".

#Noethember

A 30-day drawing challenge on the life and work of German mathematician Emmy Noether (1882–1935) (Figure 1).[d] #Noethember takes its name from German mathematician Emmy Noether and the fact that the activity took place during November 2018. This challenge was carried out by the author, Constanza Rojas-Molina, in collaboration with Dr. Katie Steckles, mathematician and science communicator based in the UK and editor of the mathematics blog *The Aperiodical*,[e] that helped promote the initiative. The project consists of one drawing each day, for a period of a month. The drawings shown in this note are taken from the 30 drawings produced by the author, unless specified otherwise.

---

[b] Mike Rohde, *The Sketchnote Handbook: The Illustrated Guide to Visual Note Taking*, Peachpit Press (Pearson Education), 2013.

[c] https://inktober.com/.

[d] https://aperiodical.com/2018/10/noethember-illustrating-a-life/.

[e] https://aperiodical.com.

Figure 1:   Cover for #Noethember by the author.

#Mathyear

A 52-week drawing challenge that explores mathematics and its interactions, done in tandem by the author, Constanza Rojas-Molina and Marlene Knoche,[f] computer scientist and illustrator based in Germany (Figure 2). The topics treated lead to the exploration of complex notions in mathematics, to reflect on inclusivity in science, and on the role of women in mathematics. The project consists of a drawing each week, for 52 weeks,[g] spanning the period of a year.

The blog *The RAGE of the Blackboard*

"Sketchnotes of Science" is part of a larger communication project that was initiated with the blog *The RAGE of the Blackboard*[h] (2014–to date), a platform where the author writes and illustrates about mathematics,

---

[f] https://www.sanguinik.de/.

[g] https://ragebb.wordpress.com/2020/03/21/mathyear-a-drawing-challenge-in-times-of-isolation/.

[h] https://ragebb.wordpress.com.

Mathematical Doodle Year
January 2019: Fractals

Week 1 (Dec 31 - Jan 6): Koch Snowflake
Week 2 (Jan 07 - Jan 13): Wacław Sierpiński
Week 3 (Jan 14 - Jan 20): Benoît Mandelbrot
Week 4 (Jan 21 - Jan 27): Fractals in Nature

Use #mathyear when posting online!
Twitter: @sanguinikDE & @Coni777
More info and details on: www.sanguinik.de

Figure 2:   First prompt of #Mathyear, logo of #mathyear by Marlene Knoche (@sanguinikDE).

women in mathematics, and the current situation of academia, according to her own personal and professional experience in Chile, France, and Germany. Traditional illustration and sketchnotes are used with the aim to interest a wide audience (see Figures 3 and 4).

Illustration and the combination of words and images are versatile tools to initiate a dialogue about mathematics and about mathematicians. Figures 3 and 4 are samples of the blog contents. Figure 3 shows a portrait of Fields Medalist Maryam Mirzakhani, including a visualization of her work, while Figure 4 is a critical depiction of an academic career in the form of the popular game Hopscotch.

In the following, we describe different aspects of the project "Sketchnotes of Science", that will hopefully allow for its implementation in other settings.

Objectives

"Sketchnotes of Science" aims to raise awareness of the central role played by mathematics in current scientific developments, and to raise awareness

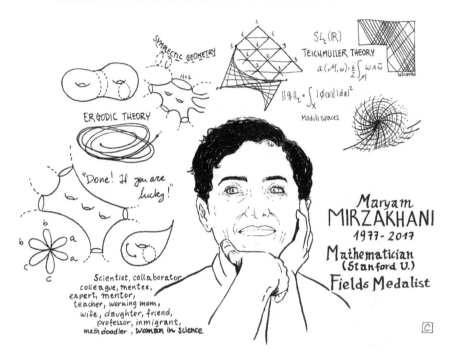

Figure 3: Portrait of Maryam Mirzakhani by the author. Blog The RAGE of the Blackboard.

of the role of women in mathematics. We expect that this will contribute to improving the perception that the lay audience has of mathematics, a traditionally unpopular subject among students, and make people more receptive to new and abstract mathematical ideas. We also expect that by exposing the audience to women mathematicians that have been overlooked in mainstream media, more women and other minorities will go into mathematics, which in turn will make academia and the mathematical society more diverse.

#Noethember

The goal of this challenge is twofold: first, to promote the work of Emmy Noether, highlighting her groundbreaking contributions to modern mathematical physics, and second, to reflect on the many difficulties she experienced in her career as a woman, in times when women were not expected to pursue university studies, and even less, scientific careers.

Figure 4: The path to professorship, collage and illustration by the author for the blog *The RAGE of the Blackboard*.

We expect to raise awareness of the role of women in mathematics and the many difficulties that women experience still today in STEM careers, which is in part responsible for the low numbers of women in academia.[i]

#Mathyear

The goal of #mathyear is to show the many disciplines where one can find mathematics, and that mathematics plays a key role in modern technology present in our everyday. With this, we aim to stimulate the audience's interest and curiosity to do further research, get interested in the topics or the people that do mathematics, and look for more information about them. Our long-term goal is for the audience to have a positive reaction to the word "mathematics" and all it entails, as opposed to evoking bad school memories or mathematical anxiety. This will, consequently, contribute to the mathematical literacy of the general audience.

Message

#Noethember

#Noethember promotes the work of Emmy Noether, who was a pioneer in mathematics and who broke the standards of her time by pursuing a university degree in mathematics and following a scientific career.

For #Noethember, a list of 30 facts ("prompts") about Emmy Noether's life and work was published in *The Aperiodical*, that were intended as inspiration for the drawings.[j] The list was curated by the author and Katie Steckles, both mathematicians. The facts combine information about Noether's life, her family and personality, the difficulties Emmy Noether had to face to pursue a career in mathematics, and facts about her mathematical research. For example:

---

[i] "A global approach to the gender gap in mathematical, computing, and natural sciences: How to measure it, how to reduce it?". In Guillopé, Colette and Roy, Marie-Françoise (eds.), *International Mathematical Union (2020)*. Project website: https://gender-gap-in-science. org.

[j] https://aperiodical.com/2018/10/noethember-daily-emmy-doodling-starts-on-thursday/.

Day 1: Emmy Noether was born on March 23, 1882 in Erlangen, Germany to Max Noether and Ida Kaufmann.

Day 2: From 1889 to 1897 Noether attended Höhere Töchter Schule in Erlangen (girls school). She studied German, arithmetic, French and English, and learned the piano.

Day 4: "As a child, Emmy gave no sign of precociousness or extraordinary ability and was indistinguishable from all the other young girls in Erlangen."

Day 7: At Erlangen, Noether was one of only two women in a university of 986 students, and was only allowed to audit classes rather than participate fully. She required the permission of individual professors whose lectures she wished to attend.

Day 9: After completing her dissertation in 1907, she worked at the Mathematical Institute of Erlangen without pay for 7 years, since at the time, women were largely excluded from academic positions.

Day 19: Much of Noether's work in abstract algebra was studying rings — sets of objects with two different ways to combine them — such as the ring of whole numbers (integers) with addition and multiplication. Of particular interest are ideals, which are particular subsets of a ring.

Day 22: Noether's (first) theorem states that every differentiable symmetry of the action of a physical system has a corresponding conservation law. It explains the mathematical origin of conservation of energy and momentum in physics.

Day 23: "If one proves the equality of two numbers A and B by showing first that A≤B and then that B≤A it is unfair; one should instead show that they are really equal by disclosing the inner ground for their equality." — Emmy Noether.

#Mathyear

The #Mathyear challenge promotes mathematics and its interactions, that is, the fact that mathematics plays an important role in many other disciplines. Marlene Knoche, computer scientist and illustrator based in Germany, inspired by #Noethember, approached the author with the idea of creating a challenge about mathematics that would span a whole year with weekly drawings. They teamed up to produce, each of them, a series of illustrations inspired by the list of prompts they curated together: a list of 52 topics gathered in 12 groups corresponding to the months of the year. In this note, we give an account of the author's experience carrying out this project.

As an example of the topics selected, the prompts for the first quarter of the year are as follows:

Month	January	February	March
Topic	"Fractals"	"Modeling"	"Cryptography"
Topics per week	W1: Koch Snowflake W2: Sierpinski W3: Benoit Mandelbrot W4: Fractals in Nature	W5: Math and Environment W6: Math and Time W7: What You Love About Math W8: Math and Sociology	W9: Cryptography W10: Prime Numbers W11: Symbols W12: Alan Turing W13: Enigma

In #Mathyear, the choice of topics is often motivated by the season or by specific dates, for example: in January it is winter in the Northern Hemisphere, therefore the selected topic is "Fractals", of which a famous example resembles a snowflake. February 14 is Saint-Valentine's Day, therefore the topic of that week is related to love, "What Do You Love About Mathematics?". We also take the opportunity to promote the role of women in mathematics. For example, September's topic is "Women in Mathematics" and November's topic is "Emmy Noether", in relation to #Noethember, which served as inspiration for this.

Figure 5:   #Noethember day 19.

Language

In both #Noethember and #Mathyear, the language used is informal. The choice of English is to reach as large an audience as possible, given that English is the standard language in mathematical research and in collaborative projects in academia. The drawings by the author for both initiatives are informal, with a strong improvisation component and with a hint of humor. This seeks to make the challenge fun and appealing to people both inside and outside of academia.

The use of the sketchnote style aims to simplify concepts and to make analogies with personal experiences or common cultural references. While we aim for the scientific accuracy of concepts, the images that illustrate the concepts are the product of a free association of ideas. For example: the notion of a "ring" is accurately defined from a mathematical perspective, and this definition follows the shape of what everyone can associate with the image of a ring, that is, a circular shape (Figure 5). Another example of free association of ideas and cultural references is given in Figure 6.

Communicator and Audience

The project "Sketchnotes of Science" is carried out by the author, a mathematician (researcher and lecturer at a public university) in collaboration with one other person who also has a scientific background.

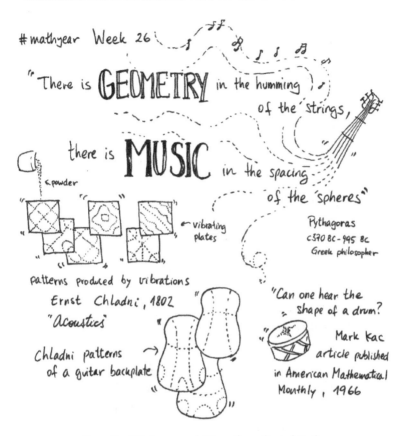

Figure 6: #Mathyear week 26, mathematics and music.

Both share tasks of promotion and engagement with the audience. The project is carried out on a voluntary basis by all parts involved, and therefore it represents the views of the communicators and not of any public institution. While both #Noethember and #Mathyear can in principle be run by one person, the advantages of sharing the tasks of audience engagement with collaborators is that it adds diversity to the opinions and relationship with the audience, it is time-efficient and it reaches a wider audience of followers in social media.

The target audience is people who are active in social media. Naturally, this implies that the audience already shows an interested in science since they "follow" topics related to mathematics. This includes science communicators, academics, teachers, undergraduate and graduate students. We expect that by reaching teachers and lecturers, with material that is appealing and free to download, we can also in the future reach pupils and students.

The project also has the potential to reach the illustration community and other artistic communities, who are permanently looking for illustration or other artistic challenges to form habits in their creative practice. This explains the popularity of artistic challenges that are seen in social media, for example #Inktober (daily ink drawings during October), #Celltember[k] (daily illustrations related to cellular biology, during September), #NaNoWriMo[l] (writing challenge throughout November).

#Noethember

Communicators: Constanza Rojas-Molina and Katie Steckles.

The author was the creator of the initiative, who participated in the curation of facts and creation of the prompt list. She was in charge of the illustrations that lead the challenge, which were set as examples and motivation for the audience to contribute their own drawings. Katie Steckles oversaw the promotion via the blog *The Aperiodical*, the curation of the facts and creation of prompt list. While this project can be run by one person alone, the project needs posts in social media and engagement with the audience on a daily basis, therefore it is useful to have at least two people who can share the responsibilities, since this can be very time consuming.

#Mathyear

Communicators: Constanza Rojas-Molina and Marlene Knoche.

The communicators curated together a list of 52 topics for each week of the year, and then set up to illustrate each topic on their own, with their distinct illustration styles. The author's choice was the sketchnote approach, following the lines of #Noethember. Both communicators promoted the challenge in Twitter and Instagram, promoted each other's work and invited other people to participate. They appealed to their respective followers, engaging with the audience on a weekly basis.

---

[k] Celltember, created by scientific illustrator Bertsy Goic, http://www.drawinscience.fr/celltember-2020/.

[l] "National Novel Writing Month", https://nanowrimo.org.

Motivation

In societies where a significant part of the population has access to the internet, people are overloaded with information. We live in the so-called Information Age, and processing large volumes of information, and being able to discern between facts and opinions is becoming increasingly important in society. In this context, images have become a powerful tool to convey information in a manner that attracts the audience and helps process complex messages. With this project, we invite the audience to reflect on complex notions of mathematics that are rendered more accessible by playful images combined with text, that is, sketchnotes. Because mathematics is a human activity, we cannot separate the abstract realm of mathematics from the human activity of doing mathematics. Therefore, we believe it is important to address in our engagement with the audience, issues that affect the mathematical community, as they affect society at large, and that are common in STEM careers: gender imbalance, lack of inclusivity, lack of professional prospects for young scientists, mobility of researchers and immigration, strongly competitive and stressful environments, etc.

Gender imbalance in the mathematical community is a pressing issue that affects the mathematical community at large. There exists a gender gap seen in mathematics at undergraduate and graduate levels that increases in the academic system, at the level of assistant professors and full professors, resulting in the almost complete absence of women at high-ranking positions where decisions about the future of academia are taken.[m] Currently, scientific societies, research and academic institutions, governments, and funding agencies are actively addressing this issue through different initiatives to close the gender gap.[n] Given this context, it is important to give visibility to women in mathematics, as a way to contribute to the variety of efforts to improve gender imbalance in mathematics. This

---

[m] As shown in the latest EU report on gender equality in research and innovation in Europe "She figures 2018", published in March 2019, https://ec.europa.eu/info/publications/she-figures-2018_en.

[n] "A global approach to the gender gap in mathematical, computing, and natural sciences: How to measure it, how to reduce it?" In Guillopé, Colette and Roy, Marie-Françoise (eds.), *International Mathematical Union (2020)*. Project website: https://gender-gap-in-science.org.

is one of the main motivations of the author, who, as a woman, Latina, and immigrant in Europe has either experienced or observed these issues herself and is willing to put them into pictures through her illustration skills.

Methods

Here we lay out the steps to create a drawing challenge around a theme.

1. Decide the main theme, regularity of the posts, rules for submission, starting date, and a catchy name for the challenge's hashtag. The use of the hashtag (#) is used to track the posts of all participants.

2. Create the list of topics, words or phrases, that will serve as an inspiration for the drawings. This is the so-called list of prompts. For a 30-day challenge like #Noethember, the list consists of 30 facts about Emmy Noether. For a year-long challenge like #Mathyear, the list consists of 52 topics. Any other combination is possible!

3. Post an announcement of the challenge in social media, well before the starting date, making the list of prompts available to the audience and inviting them to participate.

4. Materials: for the creation of the drawings, we use a limited color palette, that is, a limited number of different colors. We use easy to find, affordable materials: a sketchbook and color markers (Figure 7).

Figure 7:   Materials.

We also need a mobile phone with internet connection, or any other device that can post to social media (see Figure 8).

5. Post the sketchnotes on social media with the defined regularity (daily or weekly) using the hashtag with the name of the project (#Noethember or #Mathyear). In the same post, invite the audience to collaborate (use Figure 8 as guidelines for users).

Figure 8: User's guide. This shows the audience how to participate in the drawing challenge #Mathyear. The steps to participate in #Noethember are the same, the only difference is the topic and the regularity of the posts (daily/weekly).

It is useful to alternate the posts containing the drawings with complementary information on the topic, references, interesting articles, etc. This motivates a discussion with the audience where they can participate not only by drawing themselves, but also by sharing references and other useful information pertaining to the topic, or by asking questions. Some people might be too shy to share their drawings and prefer to participate in other ways. All forms of participation are welcome, this is not exclusive to drawings!

Time Frame

As mentioned before, the drawing challenges are set in a limited timeframe: #Noethember is a 30-day initiative. This answers to the immediacy that is inherent to these platforms, where engagement takes place in a short time window. #Mathyear is a 52-week project that aims to engage people in a longer time period.

The main idea behind challenges that involve regular activity is to create habits, and as we all know, to create or change habits is very hard. In this case, the habit to be created is that of (a) drawing and (b) thinking about mathematics. This activity takes perseverance and is a considerable time investment, therefore short-term challenges are more efficient, more focused, and have higher chances of success measured through audience engagement.

Location and Adaptability

The project is conducted online and in English to access the widest audience possible, but it is made in a way that it can be carried out in any place in the world where some kind of social media is available and in any language. Social media is a virtual space for sharing information and discussion, like notice boards or forums. In the absence of social media or internet, this project can be adapted to be fully analog, as long as there are sources of information available to participants, so that they can read and inform themselves about the topics in the prompts list.

#Noethember and #Mathyear

The use of social media is subject to time zone, local references, cultural references, and sociopolitical events during the time frame of the project.

Since the communicators are based in Europe, it is natural that in their posts and engagement with the audience, they refer to references or events taking place in their countries. #Noethember was carried out when the author was based in Germany. #Mathyear, on the other hand, was carried out when the author was based in Germany, and then in France. This influenced the tone of her illustrations and the references there.

Impact

We measure the success of the project through audience engagement: number of likes, reposts, and articles in websites specialized in science communication. We have an incomplete quantitative measurement of the impact due to the need of expertise in data analytics or the accessibility to paid platforms to analyze all the data generated.° Therefore, we rely on a qualitative impression of the impact. The illustrations by the author for both #Noethember and #Mathyear are freely available to download from her website. This encourages teachers and scientists to use them in their activities (science talks, outreach talks, articles, classroom, etc.).

Since #Mathyear is an ongoing activity, here we restrict ourselves to report on #Noethember.

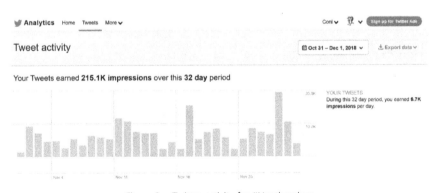

Figure 9: Twitter activity for #Noethember.

° Twitter historical data for #Mathyear and #Noethember is available upon demand.

#Noethember

Articles explaining the initiative were published in several high-impact outlets in countries outside Germany, in their respective languages[p]: in the blog of the American Mathematical Society, in the French website Images des Mathematiques (the mathematics communication platform of the Center National de Recherche Scientifique-CNRS), in the Spanish platform Mujeres con Ciencia (University of Basque Country), in the German website Scienceblogs, in the Italian website "Math is in the air". The set of 30 drawings produced by the author were translated and published in Spanish and French.[q] See Figure 10 for a selection of contributions. We note that contributor Lele Saa, a graphic artist based in London, was interested in the drawing challenge, and had no previous connection to mathematics. Her own take on all the prompts for #Noethember[r] shows what imagination and visual training can do to represent ideas without having a mathematical background in Figure 10.

The Twitter analytics histogram (Figure 9) shows the number of impressions gained on Twitter in the period during the #Noethember initiative, which is the end of October 2018 until early December 2018. We note an increase in impressions in the late part of the month when the prompts were more technical. This could be explained by the increasing promotion made by participants and the appearance of articles about the initiative in several outlets, as mentioned above. The Twitter data is available for free in the Twitter platform, however, to have a more detailed analysis of the data generated by the hashtag, #Noethember requires more sophisticated analytics platforms that are payment-based. Therefore, we have not pursued a more detailed analysis of the data so far, and considered the initiative a success given the participation of a diverse group of people and the mathematics communicators or mathematics platforms that highlighted our project.

---

[p] https://aperiodical.com/2019/01/noethember-a-retrospective/.

[q] https://mujeresconciencia.com/2018/12/07/en-homenaje-a-emmy-noether-noethember/ and https://images.math.cnrs.fr/Noethember.html?lang=fr.

[r] @lele_saa in Twitter and Instagram, see also https://twitter.com/lele_saa/status/1058024394721816579.

Figure 10: Screenshot from the article on a retrospective of #Noethember with a selection of contributions. Mathematics Blog *The Aperiodical* (www.aperiodical.com).

Concluding Remarks

We believe that the use of sketchnotes and the use of social media in the form of drawing challenges are innovative ways of public engagement in mathematics. The method used can be easily adapted to other platforms and contexts, like classrooms. The author herself has used sketchnotes with her students in the Bachelor Program Ygrec in Data Science (CY Cergy Paris University). She presented the project "Sketchnotes of Science" in the First Chilean Congress of Mathematics Communication[s] in early 2021, which had a positive response and has inspired other lecturers to do sketchnotes in their classrooms. By the end of 2021, the author expects to have feedback on these experiences.

It is important to emphasize that these drawing challenges are "Ideas, not Art", following Mike Rohde's motto. Participating in these drawing challenges is also a gentle introduction to the world of visual practice (which includes graphic recording and graphic facilitation[t]). Practicing sketchnotes for mathematical concepts can also be beneficial to communicate ideas in general. We strongly believe that developing the skill of communicating complex ideas visually (independent of the discipline) can be a useful skill in the Information Age.

---

[s] First Chilean Congress in Mathematics Communication (2021). https://sites.google.com/umce.cl/congresodivulgacionmatematica.

[t] For example, see the activities of EVP, the association of European Visual Practitioners: http://europeanvisualpractitioners.com/.

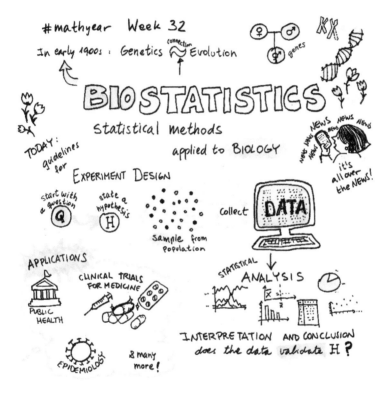

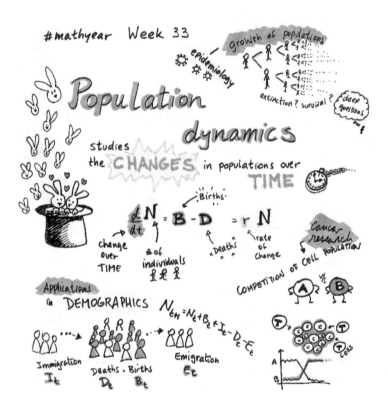

Chapter 11

Connected Curiosity

Melissa Silk* and Annette Mauer[†]

National Design Department, JMC Academy, Sydney, Australia
*msilk@jmc.edu.au
[†]amauer@myjmc.edu.au

There are many shifting ideas of how we learn and understand, and these inhabit current views supporting transdisciplinary education. Such diverse perceptions suggest a greater need now for discussions on how science, technology, engineering, and mathematics (STEM) integrate with the arts (A) across a range of learning settings, thus acknowledging STEAM. Still, experiencing STEAM is more than the acronym suggests. It is more than integration, as Keane and Keane, McAullife, and Burnard and Collucci-Gray attest [1–3]. A key aspect of STEAM learning points to the importance of considering personal and individual transformation through connected experiences [4,5], in which the dynamic contribution of dialectical emotions is appreciated and valued.

Melissa Silk and Annette Mauer are STEAMpop.[a] We have worked in a variety of learning contexts, including schools, universities, museums, and the broader community. Our reach includes local government workshops in libraries and community settings for events such as National Science Week, cross-cultural industry groups such as the Asia Pacific Architecture Forum, and festivals that celebrate the creative industries, such as Vivid Sydney. We develop integrated STEAM programs that result in innovative ways of thinking and knowing [6], and highlight the interconnectedness of such ideas and how they manifest in the world. Melissa's research with the University of Technology Sydney (UTS) explored connections between imagination, making and mathematics. Situated in a high school

---

[a] https://steampop.zone/index.html.

context over 10 years ago, Melissa collaborated with colleagues from STEM disciplines to develop a hybrid curriculum for students ranging in ages 13 through 16. Units of work were nested in a course of study called "Thinking Hyperbolically!" (exclamation included). Working with expert mathematics teacher, Jane Martin, the seed of curiosity grew into complex research projects that incorporated aesthetic practice in mathematical theory not prescribed by regulated syllabi. STEAMpop evolved from Melissa's consequent research with UTS and the relationship established with the Australian Design Centre, where Annette was the lead educator.

Together we continue to develop research-based experiences that acknowledge the relationship between science and arts society, with a particular focus on mathematics. Through our self-funded work in STEAMpop, we applaud curiosity. Many of our events and workshops are sponsored by interested parties, with a portion of funding raised through the sale of material kits for our signature projects. We engage in our work because we believe the encouragement of playful hand-making melded with mathematics promotes sense-making and revelatory experiences. Primarily in our home country of Australia, we deliver our STEAM workshops to a diverse range of audiences, including educators and students from elementary to university level, people interested in craft, tinkering and making, employees undertaking team building activities, and families out for some fun. We have created unique experiences for architects, dentists, technologists and artists, both budding and accomplished. While not mathematicians ourselves, math enthusiasts also form a large part of our audience, introducing some to new concepts of visualizing simple mathematics, while reconnecting others to familiar ideas explored in original and creative ways.

Why would we combine the knowledge areas of STEM into one acronym and add the "A" to acknowledge interconnectivity with the arts? We do this to reinforce connectedness in all its learning forms. STEM represents integrated sciences, and central to the arts are the key learning areas of humanities, language arts, dance, drama, music, visual arts, design, and new media. The STEAM acronym provides a neat catchall title for creative learning ecologies, where cultures of thinking, creative programming, inventive teaching, and learning rely on play, curiosity, fearlessness, passion, and purpose [4,7]. Our practice-based research and actions in balanced STEAM convergence have shown how these emotional inputs incite memorable experiences [6]. Our work in STEAMpop investigates how blending and embedding the arts in STEM emphasizes the value of STEAM and is a powerful method of enacting authentic transdisciplinary learning that is accessible for all types of learners, not only teachers and students.

Playful Engagement Through Making with Mathematics

Our projects and workshops provide a playful splash of creativity in the vast sea of transdisciplinary learning potential. Here, we are referring to math-making experiences that never lose sight of the complexity, beauty, and elegance of their mathematical foundations. Most people with reasonable manual dexterity, generally 8 years and older, with a healthy curiosity and a measure of persistence and grit, are able to participate in our workshops [6]. While an ability to visually recognize patterns is desirable, we have experience assisting participants with vision impairment, using the sense of touch to demonstrate the physical aspect of 3D tessellations.

A range of mathematical content is explored through physical activities related to making. Maeda argues, there is no greater integrity, no greater goal achieved, than an idea articulately expressed through something made with your hands [8]. Endorsing Maeda's approach, our research philosophy emphasizes individual freedom to engage with mathematical concepts through making, by asking makers not only to make, but to *think about* and *feel* the mathematics underpinning their actions of making "in the moment" (Figure 1). Dewey describes such immersion as "Erlebnis" [9]. For instance, the perceived simplicity in making a "Binary Bug" (Figure 2), reveals deep conceptual connections between the knowledge systems of biomimicry, binary, elementary geometry, nonlinear relationships, crystallography, patterning, engineering, and the elements and principles of design. These connections are revealed via initial presentations and questioning, and reinforced while participants make their Bugs, and are explained in more detail later in this chapter. This is not reductive learning, but rather, transformative constructivism, hybridized with the beauty of

Figure 1:   Lumifold at Momath 2015.

Figure 2:   Binary Bug sample.

Consilience Theory [10], which describes life and knowledge systems in terms of how *everything connects.*

Playfulness inherent in math-making teases the resistance out of those who refer to themselves as not mathematically minded. While for some, using one's hands results in the application of what Dweck refers to as a "growth mindset", and for others, is evidence of the fact that STEAM capabilities are greatly enhanced by both rational and non-rational elements of consciousness [11–13]. In saying this, we acknowledge that permission to physically play, make, achieve, and fail, generally encourages people to

Figure 3:   Hungry Birds at Moves 2019.

step out of perceived comfort zones bounded by words and equations (see Figure 3). Operating inside a theory of how everything connects results in the synthesis of cause and effect connections that situate STEAM learners continuously "on the breach" [1]. It's an interesting emotional state in which to find oneself. "Oh my goodness, I get it!" (Participant, 2017).

Harnessing the Power of Connection

The projects we develop for delivery within education, community and corporate environments share one clear purpose; that is, to build intercultural capacity by harnessing the power of visual and creative arts as a strategy for understanding STEM concepts, particularly mathematics. Beyond the STEAM acronym, we believe in asking "why" and "what if" regularly. In terms of raising public awareness, understanding, or mathematical literacy, defined as a process of creative and critical thinking that encourages acceptance and openness with a view to changing basic existing paradigms constructed from our attitudes and behaviors [2,4,14]. Hence, making continual and implicit STEM connections within an Arts context and vice versa, demonstrates how transdisciplinary learning shapes the development of personal and professional identities [15]. We see math-making as a great leveler, unhinged from subject specific skill or expertise, and let loose amongst what is already known, and what can be discovered. Therefore, the element of risk inherent in undertaking any

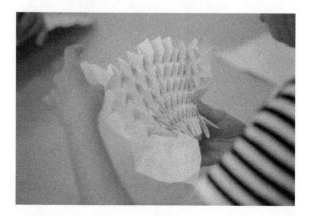

Figure 4:   Lumifold 2016, Brisbane.

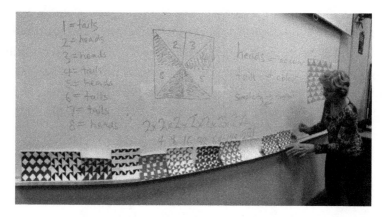

Figure 5:   Binary Bugs teacher in classroom.

of our projects is important. Through engaging in "learning by doing", people of all ages, and not only students or teachers, in our experience, appreciate the wonder of finding connections. For example, Figures 4 and 5 show integrated learning related to concepts of geometry and biomimcry (Figure 4), and binary, probability, and art/design (Figure 5). Engagement with the interrelatedness of math concepts and art/design practices allows for a common thread to emerge and weave its way through concretion to abstraction in a continual loop. This can be transformative for participants who come to recognize the connections between the mathematical science underpinning a particular "making" experience, and the aesthetic experience itself.

Expanding Curiosity

Curiosity drives STEAM learning in terms of knowledge transactions and connections, haptic engagement, risk-taking and imagination. Aligned with Nobel Prize winner Richard Feynman's view, the pleasure of finding things out acknowledges the balanced aesthetic inputs from both artist and scientist. As former Australian Chief Scientist suggests, "no clever country would encourage its most STEM-literate people to pursue only traditional research paths" [16]. Finkel goes on to say that his own experience would reveal that he found opportunities in unexpected places.

Embedding the arts in STEM affords disruption of traditional modes of learning by fully integrating disparate content in ways that are imaginative, challenging, and relate to the real world. Some of our work in schools has shown that expanding teachers' curiosity for STEAM afforded them new levels of confidence in risking transdisciplinary approaches to curriculum-bound pedagogy. Similarly, in our public engagement activities the "wow" or "aha" moments abound, as connections between STEM disciplines and the hand-making of aesthetic objects are brought to light (Figures 6 and 7). These connections often elicit questions and further explorations. The action of making is central to the notion of embedding science creativities into the meaning of a lifelong education where getting your hands dirty

Figure 6:   Asia Pacific Architecture Forum 2016.

Figure 7: Mobifold, Australian Design Centre, 2020.

is crucial to understanding "why you made what you made, and owning the impact of that work in the world [8]".

Mindful Making

We consider engaging in STEAM experiences as visceral. Finding oneself immersed in our STEAM experiences relies on maintaining curiosity and perseverance, with acute awareness of how these states are embodied as feelings. The act of folding offers moments of contemplation and meditation through the challenge and joy of making. Ackerman has said "deep play is the ecstatic form of play" [17] (p. 12), and at its peak, all elements are visible and intense. Not unlike the personal aesthetic experience viewed by Hirsh-Pasek *et al.*, Robinson, and Csíkszentmihályí, as an experience of flow, in which a person loses a sense of time while completely engaged in an activity [18–20]. Hence, the outcomes of our STEAM experiences are aligned not only with a body of knowledge related to transdisciplinarity, but also with studies of multiple creativities and the inseparable link between neuropsychology and physical activity [3,12,21–23].

The literature on play has foregrounded exploration and discovery as central to the notion of play. To be swept up in a deep state of play, immersed, engaged, oblivious to the surrounding environment, incites

Figure 8:   STEAMpop's immersive experiences.

feelings of balance, focus, creativity, challenge, and possibility [17,24–26]. It is for these reasons that our STEAM experiences, despite inciting dialectical emotional responses, can be categorized as the interrelationship between challenge, immersion, deep play and mindfulness (Figure 8).

Broadening Audiences, Welcoming Everyone

When you investigate mathematics through the arts, discovering connections between knowledge areas and learning demonstrate content interrelationships more explicitly, and are often a lot of fun. Our math-making experiences target a range of groups. Our audience is varied, ranging from educators through professional associations, accreditation bodies, and via request from individual institutions. Our math-making workshops contribute to the outreach activities at UTS, crossing faculties of Science, Design, Education and Transdisciplinary Innovation. In schools, we provide professional learning sessions that are accredited by the New South Wales Educational Standards Authority (NESA). We believe our greatest reach, however, is related to the public events we present through collaboration with museums and community organizations such as MoMath, the National Museum of Mathematics (New York City), Vivid Sydney, the Museum of Applied Arts and Sciences (MAAS), and local/regional Australian galleries.

The consistency of our designed experiences finds us using mathematical paper folding (origami sekkei) as a foundation for many of the projects presented to schools, tertiary institutions, and the general public. Origami sekkei, or technical origami, relies on the creation of crease patterns derived prior to folding being based on mathematics. It allows for the creation of extremely complex multi-limbed models to be formed [27]. Building on the

Figure 9: STEAMpop experiences — Lumifold and Circulus.

duality of simple/complex making associated with origami sekkei, our projects incorporate one or more STEM or arts perspectives such as mechatronics, robotics, photoluminescence, body adornment, augmented reality, and the creation of digital visual narratives. There is a constant dialogue between ourselves as facilitators, and collaborators such as Australian Indigenous scientists, academics, local artists, teachers, and students. All participants in our workshops and events provide valuable feedback, allowing the evolution of our research and practice, which is based on continued contribution to building innovative cultures of thinking (Figure 9).

Workshop Experiences

Lumifold

> It's so good to do something different and playful together.

Our signature workshop experience is Lumifold. The Lumifold experience involves making paper lamps utilizing origami sekkei techniques as a way of integrating art, design and STEM. It promotes ideas related to biomimicry, elementary symmetries, glide reflection, and translations while learning about structure, strength and stability. Here, we explore the geometry of the crease pattern in terms of where the lines intersect, what angles they form, and in what direction the creases fold: are they valley creases or mountain creases?

Lumifold is accredited as a course by NESA for 2 hours of Teacher Professional Learning and is delivered to teachers in early secondary and upper primary schools in NSW, Australia. Teachers gain an understanding of terminology used in paper engineering and in particular, the "glide reflection" technique traditionally used in the periodic water-bomb fold. In Lumifold, participants discover how flat-to-form concepts can be realized and connected to forms in nature. By flat-to-form, we mean transforming two dimensional shapes into three-dimensional forms as seen in Figure 12. The activity provides opportunities for the recognition and discussion of numerous mathematical and STEAM concepts such as tessellations, randomness, geometry, symmetries, biomimicry, flexibility, and visual complexity, while making an auxetic structure. Auxetics is a representation of the so-called negative Poisson's ratio, wherein materials can be expanded in two directions at the same time. Amongst other applications, this is used currently in biomedical research.

During the making component of the Lumifold experience, participants fold paper templates of varying sizes into hills or valleys (up or down) according to origami sekkei rules and conventions; red indicating mountains and green indicating valleys (Figures 10 and 11). The mathematics in Lumifold includes Maekawa's Theorem: at every vertex where creases intersect in a flat origami crease pattern, the difference between the number of mountain and valley creases is always two. When folded, the paper no longer retains its two-dimensional integrity and begins to curve. There are two specific patterns that we have developed called Lumiball (the flexible spherical structure seen in Figure 12 left) and Lumitall (the rigid cylindrical structure seen in Figure 12 middle).

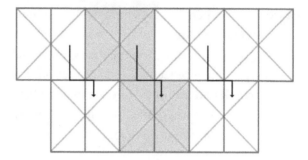

Figure 10:   Lumitall tessellation.

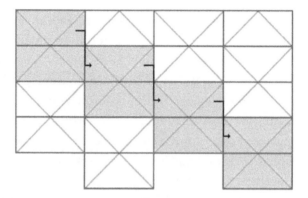

Figure 11:   Lumiball tessellation.

Figure 12:   Lumifold lamp variations and illumination.

Lumifold experiences have been presented to a broad range of audiences outside the school context. Community engagement has been facilitated in large festivals such as *Vivid Sydney, Light, Music and Ideas Festival*, *Sydney Craft Week*, and *UTS Spark Festival*, and in corporate settings like Google. The math concepts are reiterated while the participants

Figure 13:   Lumifold's wondrous geometry.

are in the process of making the lamps so that the connections between what they hold in their hands has a direct relationship to the wondrous geometry. On completion of the paper lamps, we illuminate them by placing LED bulbs inside a correspondingly folded biomimetic cup. The room lights are dimmed and the lamps are lit. The final reveal gives rise to gasps of amazement as each lamp becomes a glowing beacon throwing up colored star shapes on the ceiling, demonstrating another unexpected dimension of this experience (Figure 13). This surprising consequence continues to evoke joy and delight from participants as they further explore the geometry.

Binary Bugs

I didn't think I could do this… I'm so proud of myself!

During our own experiments in developing Lumifold we discovered a myriad of ways of utilizing the same mathematical fold. One of our discoveries led to the emergence of Binary Bugs (BB), another of our endorsed teacher professional learning courses. BB has proven the most accessible for schools as it compresses the construction of knowledge, skill and experience into the creation of a small aesthetic artefact that delights students.

BB developed as a method of exploring elementary symmetries in mathematics, plus the creating of patterns resulting from combining ideas related to probability and binary. Drawing on the "Law of Large Numbers" [28] in which large sequences of coin tosses will result in about equal heads and tails, the skewed splits that form the BB patterns are due to the small number of coin tosses; eight to be precise. The surface pattern applied to

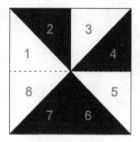

Figure 14: Binary pattern created by randomly tossing a coin eight times: Sequence = heads, tails, heads, tails, heads, tails, tails, heads.

Figure 15: Application of binary patterning.

the BB template paper is created by tossing a coin eight times to determine a 2D black and white (or color/no color) design on a small square single unit template (Figure 14). The binary sequence represented in the single unit is then transferred to a larger paper template and tessellated across its surface.

Constraining the pattern construction by restricting the start position of the sequence results in a possible 256 different designs which is immediately relatable to the base-2 numeral system, known as binary. Typically ones and zeros, binary numbers are represented in the BB experience by color/no color tessellations that are further enhanced by the introduction of the geometry of the glide reflection pattern (see Figure 10). The sequence of images depicted in Figure 15 shows the application of a binary-related pattern onto the surface of a pre-scored paper template indicating the geometry related to folding the "glide reflection" pattern. Instructions for creating the patterns and folding the bug-like shape are available from STEAMpop.zone on request. The patterned paper is then folded into a biomimetic bug shape, guided by scored lines embedded into its surface. Figure 15 displays a range of patterned shapes not yet fixed to the backing board or illuminated by LEDs. The board, or base, is designed to house a small LED (light) which remains accessible for switching on and off when the bug shape is attached. "Feelers" are optional.

The BB experience explores the complexity generated by the interaction of two simple systems; a randomly created two-dimensional binary pattern (based on probability) and the structure of three-dimensional paper folding according to origami sekkei rules. The geometry of the 3D pattern embedded in the paper is enhanced by coin tossing to determine a 2D black and white (or color/no color) design. Hence the idea of binary merged with the mathematics of probability (Figure 15). Our explorations in patterning are reminiscent of Truchet tiling and the work of art theorist Ernst Gombrich, *The Sense of Order*, a text detailing the psychology of decorative art [29]. The questions arising from applying a binary pattern to the paper surface have incited discussion related to base-2 numbering systems and where such systems exist in the real world. For young people, this is an interesting corollary with their own digital lives.

Introducing the concept of biomimicry to the BB experience adds to the relevance of finding connections between math and nature. An interesting aspect of exploring the glide reflection sequence through a biomimetic lens affords the makers new understandings of how transformations on the coordinate plane can result in 3D forms from 2D shapes (Figure 16). A range of bug sizes and shapes can be made using the templates, based on the choice of variables in grid numbers and the size of the initial single unit square (Figure 17).

The BB experience provides a starting point for further STEM investigations. The bugs can be illuminated using either ready-made LEDs or by creating electrical circuitry requiring the construction of an interactive switch (see Figure 17 right). Some have used Arduino and other

Figure 16: Binary pattern making and flat to form experiences in Binary Bugs.

Figure 17: Binary patterns unfolded/folded with paper circuitry (on right).

programmable hardware to create different light sequences. Timing and sensor activation are another learning extension possibility. There are a number of ways to imagine and produce immersive installations using the bugs, to create a more expansive demonstration of the binary sequences. The application of surface decoration can also lead to discussions of the role of pattern and ornamentation across cultures, pattern recognition, and its role in perception, and the aesthetic appeal of pattern applied to decorative arts and fashion industries. Such possibilities highlight the connections with mathematical theory including an introduction to plane crystallographic or wallpaper groups.

Hungry Birds

So I'm also learning while playing around.

Hungry Birds (HB) provides a playful metaphorical experience not only for the math-minded and creative but those drawn from different disciplines such as engineering, IT, design, fine arts and crafts, and education. The HB experience includes a brief exploration of mathematical curves through Cartesian plotting and graphing and conic sections, as well as terminology related to art and design principles. The experience also incorporates introductory mechatronics and coding.

Here, we engage with paper engineering to construct a mathematical model representing a hyperbolic paraboloid (HP), as shown in Figure 18 (left). The HP is a shape that, amongst a myriad of applications, also remarkably resembles the beak of a bird! (Figure 18). The beaks represent

Figure 18: Hyperbolic paraboloid beaks and "motherbird".

Figure 19: Constructing the mini-nests in Hungry Birds.

our concessional link with biomimicry, reinforcing another of STEAM's potent learning concepts. We explore the concept of biomimicry in folded mechanisms by experimenting with articulated movement of the 3D forms such as the opening and closing of a beak! Ergo in HB, we aim to place humor at the front and center of embodied STEAM learning. While the HB experience focuses on innovative and cohesive integration of art and design with STEM theory and concepts, it is primarily directed toward physically visualizing mathematics in an accessible, achievable, and entertaining contextual format.

Participants in the HB experience make as many HP paper structures as possible, one of which is contributed to the collective construction of the "nest", sometimes protected by a watchful "motherbird" (see Figure 18, right). After folding, the HPs are attached to specific pre-made mechanical components which we call the "mini-nests" (Figure 19), which contribute to the larger collective installation (see Figure 20 left). The development of the mini-nests must be attributed to the collaboration with Corey Stewart, from CJS Robotics [30]. Once HP attachment is achieved, the

Figure 20: Activating the nests and utilizing leftover HPs.

motion sensors built into the mini-nests are activated and tested. At this stage, opportunities exist in manipulating aspects of the HB movement via participant programming/coding.

There are several scalable outcomes relatable to the HB experience. The collective structures can be simply realized as a mini-nest, more literally described by Stewart as "an electronic, Arduino-controlled, ultrasonic object detection, stepper-motor-driven, expanding and retracting, led lit screw mechanism!". HPs not included in the collective nest can be utilized in playful and creative problem-solving activities leading to the formation of assorted polyhedra (Figure 20).

The coupling of the HP folding experience along with the robotic-like nature of the mini-nest system gives rise to discussion of many emerging technologies such as soft robotics, robotic end-effectors and manipulators, social robotics, ethical use of automation, and the sustainability of ongoing technological advancement. Figure 19 displays the design for the HB mini-nests and Figure 20 illustrates the realization of the design in the HB experience. Follow the footnote link to an example of the HB experience (pre-COVID-19). This shows a group of eighty participants contributing to the formation of a giant nest.[b]

---

[b] https://www.youtube.com/watch?v=jXQeph-tHI0.

Circulus

I think I was curious throughout the entire time... I just want to know and do more of this stuff!

Our Circulus experience journey began in a collaboration with paper artist Lisa Giles.[c] Lisa has folded origami structures for more than 10 years, drawn to investigating how origami influences the design of artist's books while completing her Master of Arts study. Like most people, Lisa started folding square papers, generally considered the basis of traditional origami structures. During her own explorations she stumbled upon "Whatman" filter papers, typically used in the science laboratory, and was able to access circular papers in many sizes.

After multiple experiments, Lisa and STEAMpop arrived at a folding activity that is a variation of modular origami, where multiple units are assembled to create a larger, more complex form. Each individual piece is a circular version of a "Turkish Map Fold" which can be categorized as "shape-memory origami". Its mechanical behavior is reliant on the properties of material memory such that, after unfolding, the paper structure can be easily refolded and returned to its compact shape. This method of folding allows the paper to be opened and closed in one swift motion, like a map. Traditionally, this map fold is constructed from a square or rectangular base using inclined folding lines, an essential element of a traditional Miura fold [31]. Our point of difference is that we replace rectangles with circles, referencing the research of psychologists Bar and Neta, who found that on average, people prefer curves [32]. Our initial garland comprised circles of equal size. However, differentiating the size of the circular papers resulted in a surprising new discovery: a flexible ammonite-like form that you can wear!

An important aim of this folding and constructing experience is to encourage exploration of math in order to understand its relevance to the creation of the aesthetic product. We consider the aesthetic product to be of equal importance to the aesthetic experience. The use of circles

[c] http://www.lisagiles.com.au/.

Figure 21:   Model used for calculating the area of a sector.

presented us with the opportunity to introduce geometry, including the relationship with Pi, quadrants, sectors, segments, and radians. The model we use to explain this uses paper and string to provide a visible learning experience leading to discussion of formulas for calculating the area of a sector (Figure 21). Of particular relevance is Kawasaki's Theorem in the mathematics of paper folding, and its applications in art, design, and industry. This theorem denotes the sum of angles to be folded around a vertex will be exactly 180 degrees when the paper is in a flat-fold state. The beautiful, wearable object we create relates to biomimicry in design and is reminiscent of a vertebral column and its articulation, and a Nautilus shell (Ammonites), a symbol of proportional perfection.

Participants in the Circulus experience are issued a set of paper circles of incremental sizes from around 30mm to 60mm diameter. Two circles are folded then glued at a distinct angle from one another (see [33]). The process is repeated until a naturally curving structure emerges. We use a variety of papers, repurposing old maps and art books as well as color papers to enhance the aesthetic and creative experience. Repurposing materials allows the participant freedom to experiment with color and pattern, or express emotional connections to place and memories (Figures 22 and 23).

Figure 22: Participating in Circulus.

Figure 23: Forms constructed in Circulus.

In Progress and Future Directions

Mobifold

STEAMpop and creative collaborator Lisa Giles continue to explore the beauty of mathematical ideas and forms in paper. More recently, we have designed STEAM experiences providing the opportunity for participants to play with the mysterious Möbius strip. Drawing on the explorations and artworks of Escher, Jun Mitani, and Erik Demaine, the Mobifold experience explores two strands of experimentation and play. The first is related to twisting paper templates of various widths into Möbius strips, resulting

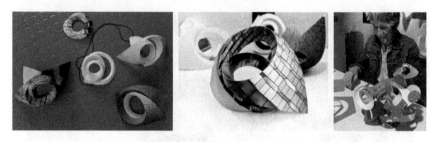

Figure 24: Experimenting with form in Mobifold.

in items for body adornment or display as small sculptures. The second is combinatorial folding of curved shapes using pre-scored paper templates including pentagons, triangles, circles, and squares in relationship with a central circular cavity. The templates can be folded in various ways to create small aesthetic sculptural objects (Figure 24). All the shapes and strips are cut from a single sheet of paper, some of which are printed with beautiful patterns. This flat-to-form experience is a playful exploration of surface, form, and space.

Flextales

Our Flextales experience begins with storytelling and the manipulation of a sequential set of photographic images positioned within four hexagons, then deconstructed in order to be reapplied to a secondary 2D paper template. When constructed, the template results in a 3D, four-sided, geometric, rotating hexagonal shape. In this way our approach is slightly different to the standard method applied in mathematical making. Much of the discussion arising in this experience is related to comparing the physical properties of the hexagonal units made with equilateral or isosceles triangles. Mapping digital photographic images onto positional templates before printing and constructing poses rich mathematical questions as well as literacy components in terms of narrative creation, or how to tell a visual story.

The characteristics of the hidden geometries inherent in Flextales can be perplexing for their makers. Breaking the original hexagons apart into triangles and arranging the triangles on the purpose-designed template is messy, but fulfilling, when the images are re-set into a "readable" image

Figure 25:   Flextales construction.

Figure 26:   Lumicrux geometries.

in the story sequence. The shape is manipulated, or "flexed", to reveal the story while rotating from one hexagonal face to the next (Figure 25). The single artifact is in itself, quite compelling, and when recorded in flex motion alongside many others, the stories have a powerful emotional effect.

Lumicrux (under southern stars)

In the spirit of sustainability our current STEAM experience makes use of discarded or supernumerary "glide reflection" paper templates. Piloted pre-COVID-19 in 2020 at MoMath in New York, Lumicrux participants used mathematics and folding to create beautiful photoluminescent forms — paper stars that glow (Figure 26). The stars and their makers became part of a special dance emulating the tiny floating phytoplankton of the Southern Ocean before resting in the position of Acrux, Becrux, Gacrux, and Delta Crucis. These are the four large stars of the constellation Crux, commonly known (in English) as the Southern Cross, part of Australia's Indigenous cosmology for millennia. Lumicrux is a speculative math-making experience in which the audience transforms itself into a representation of the constellation. The Lumicrux STEAM experience sheds light on intercultural and scientific understandings while also providing magical making moments.

Making an Impact

We are excited by the fact that the fusion of mathematical concepts and arts/design practices creates an unexpected but welcome divergence from solid subject specific knowledge, or known knowns. Our work encourages people to explore new ways of understanding STEM concepts as well as developing novel creative approaches to visualizing, enacting, or embodying such concepts. For elementary or high school educators, who teach students between 11 and 18 years of age, transdisciplinary learning warrants the incorporation of relational understandings in the subjects they teach, as a means to challenge convention. STEAM asks those teachers to step outside the comfort of personal and professional traits, irrespective of the practice of teaching remaining siloed and bound by seemingly rigid curriculum parameters. Hence, teachers participating in STEAM learning might experience heightened emotions when operating in unfamiliar knowledge or skill territory. Our work has shown that it is within these intense learning moments that STEAM has a transformative capacity. Such capacity is transferable to a general audience, where the impact of connected curiosity generated by our STEAM experiences has resulted in newfound interest in mathematics through expressions of delight and joy.

References

[1]   Keane, L. and Keane, M. (2016). "STEAM by design". *Design and Technology Education: An International Journal,* 21(1), 61–82.

[2]   McAuliffe, M. (2016). "The potential benefits of divergent thinking and metacognitive skills in STEAM learning: A discussion paper". *International Journal of Innovation, Creativity and Change,* 2(3). www.ijicc.net.

[3]   Burnard, P. and Colucci-Gray, L. (eds.) (2020). *Why Science and Art Creativities Matter.* Leiden-Boston: Brill/Sense.

[4]   Ritchhart, R. (2015). *Creating Cultures of Thinking: The 8 Forces We Must Master to Truly Transform Our Schools.* San Francisco: Jossey-Bass.

[5]   Dweck, C. S. (2008). *Mindset: The New Psychology of Success.* Random House USA Inc.

[6] Silk, M. (2020). *The Value of Me in STEAM*. (Doctor of Philosophy). University of Technology Sydney, Sydney.

[7] Wagner, T. (2012). *Creating Innovators — The Making of Young People Who Will Change the World*. New York: Scribner.

[8] Maeda, J. (2012). "STEM to STEAM: Art in K-12 is key to building a strong economy". *Edutopia*. http://www.edutopia.org/blog/stem-to-steam-strengthens-economy-john-maeda.

[9] Dewey, J. (1938). *Experience and Education*. New York: Collier Macmillan.

[10] Wilson, E. O. (1999). *Consilience, the Unity of Knowledge*. New York: Vintage Books.

[11] Csikszentmihalyi, M. and Robinson, R. E. (1990). *The Art of Seeing*. Los Angeles, California: Getty Publications.

[12] Pallasmaa, J. (2009). *The Thinking Hand*. United Kingdom: John Wiley & Sons Ltd.

[13] Hanney, R. (2018). "Doing, being, becoming: A historical appraisal of the modalities of project-based learning". *Teaching in Higher Education*, 23(6), 769–783. Doi:10.1080/13562517.2017.1421628.

[14] Schleicher, A. (2018). "Reimagining the teaching profession". *Teacher Magazine*. https://www.teachermagazine.com.au/columnists/andreas-schleicher/reimagining-the-teaching-profession?utm_source=CM&utm_medium=bulletin&utm_conten%E2%80%A6.

[15] English, L. D. (2016). "STEM education K-12: Perspectives on integration". *International Journal of STEM Education*, 3(3). Doi:https://doi.org/10.1186/s40594-016-0036-1.

[16] Finkel, A. (2016). *Australia's STEM Workforce: Science, Technology, Engineering and Mathematics*. Canberra: Office of the Chief Scientist, Australian Government. http://www.chiefscientist.gov.au/2016/03/report-australias-stem-workforce/.

[17] Ackerman, D. (2000). *Deep Play*. New York: Vintage Books.

[18] Hirsh-Pasek, K., Zosh, J. M., Michnick Golinkoff, R., Gray, J. H., Robb, M. B., and Kaufman, J. (2015). "Putting education in "educational" apps: Lessons from the science of learning". *Psychological Science in the Public Interest*, 16(1), 3–34. Doi:https://doi.org/10.1177/1529100615569721.

[19] Robinson, K. (Producer). (November 2010). "Changing education paradigms". http://www.youtube.com/watch?v=zDZFcDGpL4U.

[20] Csíkszentmihályí, M. (1990). *Flow, the Psychology of Optimal Experience* (2nd edn.). New York: Harper Perennial.

[21] Fenyvesi, K., Lehto, S., Brownell, C., Nasiakou, L., Lavicza, Z., and Kosola, R. (2020). "Learning mathematical concepts as a whole-body experience: Connecting multiple intelligences, creativities and embodiments within the STEAM framework". In L. Colluci-Gray and P. Burnard (eds.), *Why Science and Art Creativities Matter* (pp. 301–336). Leiden-Boston: Brill/Sense.

[22] Fiorilli, C., Gabola, P., Pepe, A., Meylan, N., Curchod-Ruedi, D., Albanese, O., and Doudin, P.-A. (2015). "The effect of teachers' emotional intensity and social support on burnout syndrome. A comparison between Italy and Switzerland". *Revue Européenne de Psychologie Appliquée*, 65, 275–283. Doi:https://doi.org/10.1016/j.erap.2015.10.003.

[23] Gulliksen, M. S. (2016). "Embodied making, creative cognition and memory". *FORMakademisk,* 9(1). Doi:https://doi.org/10.7577/formakademisk.1487.

[24] Burnard, P., Dragovic, T., Jasilek, S., Biddulph, J., Rolls, L., Durning, A., and Fenyvesi, K. (2018). "The art of co-creating arts-based possibility spaces for fostering STEAM practices in primary education". In T. Chemi and X. Du (eds.), *Arts-based Methods in Education Around the World* (pp. 247–279). River Publishers Series in Innovation and Change in Education.

[25] Craft, A. (2015). *Creativity, Education and Society*. London: Institute of Education Press.

[26] Holdener, J. (2016). "Immersion in mathematics". Paper presented at the *Bridges Art Math Conference*, Jyväskylä, Finland.

[27] Hull, T. (2015). "Origami: Mathematics in creasing". *The Conversation.* https://theconversation.com/origami-mathematics-in-creasing-33968.

[28] Lachowska, A. (2016). *The Math of a Coin Flip.* Yale University Press Blog. https://yalebooks.yale.edu/2016/02/03/the-math-of-a-coin-flip/.

[29] Gombrich, E. H. (1992) *The Sense of Order. A Study in the Psychology of Decorative Art.* First published 1979. London: The Phaidon Press Ltd.

[30] Stewart, C. (2018). "CJS Robotics". https://www.cjsrobotics.com/.

[31] Angsüsser, S. (2012). "Map folding techniques in the digital age". https://icaci.org/files/documents/ICC_proceedings/ICC2013/_extendedAbstract/431_proceeding.pdf.

[32] Bar, M. and Neta, M. (2006). "Humans prefer curved visual objects". *Journal of Psychological Science*, 17(8).

[33] Mauer, A., Silk, M., and Giles, L. (2019). "A synthesis of sectors". *Bridges 2019 Conference Proceeding.* Linz, Austria. http://archive.bridgesmathart.org/2019/bridges2019-605.pdf.

Chapter 12

My Experience of Producing Mathematical Films

Ekaterina Eremenko

Institut fur Mathematik, TU Berlin, Berlin, Germany
ee@eefilms.de

In this chapter, I would like to share my experience in making films about mathematics, about the tasks I set myself, about the difficulties I encountered, and about the reactions of ordinary spectators, and professional mathematicians.

In this article, I want to put the study of purely educational films beyond the scope of this paper. I am interested in representations of mathematics in popular films aimed at a wide audience, whose viewers are not only going to gain specific mathematical knowledge but enjoy the film as well.

Colors of Math, 2012, 60 minutes

This project began in 2011, when my friend and classmate in the Faculty of Mechanics and Mathematics at Moscow State University, Yuri Tschinkel, asked me to produce something cinematic that he could show at his conference presentations to non-mathematical scientists. He also complained to me that chemists at popular lectures might blow something up, physicists might show some of their experiments, while mathematicians' work is not as dramatic.

The main problem in making a film about mathematics I have formulated for myself is this: How do we show in film such an abstract activity such as mathematics? Cinema is first and foremost a visual art.

In mathematics, sometimes the correct formulation of a problem is already the halfway point to finding its solution. The language and structure of my film were therefore an attempt to solve this problem.

I suggested that I could organize the film to showcase the difficulty of visualizing mathematics through human senses.

I decided to show several mathematicians working in different countries and in different fields, and to project the work of each of them onto one of the human senses of the world — taste, sight, smell, touch, hearing, and sense of balance. I didn't take this kind of projection too seriously. I know there are many serious scientific works, for example, about the relationship between mathematics and music. But for me this idea of showing mathematics through the senses was just a game, a kind of skeleton for the film, a kind of tower which I built throughout the film, and then in the last chapter, treating it ironically in a way, I destroyed it.

But it was a game that serious mathematicians agreed to play with me and for that I am immensely grateful.

This is where I want to stop and talk a little about working with them.

They are Cédric Villani (taste), Anatoly Fomenko (sight), Aaditya V. Rangan (smell), Günter M. Ziegler (touch), Jean-Michel Bismut (sound), and Maxim Kontsevich (sense of balance).

The subjects we chose had to be from truly contemporary research, with all the mathematicians talking about what they were really thinking and working on at the time. However, I decided to exclude professional vocabulary from the film so that it does not become unintelligible to the audience.

With each character I negotiated rough content in advance. Yuri Tschinkel was very involved and helped me by giving me a broad idea of what each character was doing. One of the best cameramen in Russia, Pavel Kostomarov, agreed to work on this film with me. I regard him as a collaborator on the film, along with the composer Mike Schroeder.

Pavel and I were very careful in choosing, and when we had the opportunity, preparing the locations for the shoot.

What was not quite usual for documentary filmmaking was that I almost refused to film my characters in their usual domestic surroundings, with people close to them and their families. I deliberately decided not to touch on their personal lives at all and to talk only about their professional lives. We decided to show them alone with their main activity: mathematics.

The first character I went to see about filming was Fields laureate Cédric Villani, then director of the Henri Poincaré Institute. I wanted him to present mathematics in connection with taste. Cédric sat at his director's desk in front of a pile of documents, took one paper at a time from the pile, slapped a stamp on it, and set it aside with satisfaction as he moved on to the next paper. It was the epitome of a bureaucrat's job, and when I entered, I could not help laughing. To which he confirmed without the slightest embarrassment that he enjoyed putting stamps on documents because it symbolized the completion of an action, unlike mathematics, which never ends. "Solving one problem immediately leads to the next," said Cédric. I offered him a part in my film, and briefly outlined my idea — to use a dubious parallel with the senses. I specified that I'd like him to explain to me what he does and for what work he got a Fields Medal.

There are sometimes surprisingly generous people among mathematicians. Cédric Villani was a stranger to all kinds of arrogance. A few weeks later Cédric emailed me to say he had been invited to judge a pastry making competition in Lyon. "It all adds up," Cedric wrote. "They need a pastry contest, you need cooking for a film, let's shoot the story in Lyon!"

The competition was for pupils of the École Normale Supérieure, which Cédric oversees and tries in every way to support and promote. Lyon is also considered the center of French cuisine. I felt the wind was blowing at my back in this bizarre undertaking and we agreed to a shoot.

We spent a few days around Cédric and it was fascinating to see how Cédric can use different objects and situations to talk about mathematical ideas and see mathematics where other people don't.

At the baking competition we witnessed Cédric, with his life's excitement, making sure his favorite entries were recognized and awarded winning prizes. Cédric mentioned that there are various ways of distributing the prizes and counting the votes of the jury, which can make the final results differ greatly.

Then Cédric had wire figures prepared for the upcoming festival of lights. He used them to illustrate that something may seem like chaos, but if you change the angle you can see order and detect structure. Cedric used teacups in the café, and later, complex Lyon dishes at dinner to talk about entropy and the Boltzmann's equation.

A blackboard with formulas was also used in the film, but adhering to my decision not to frighten the audience with incomprehensible words and terms, I showed these formulas briefly in a fast-changing slide show. However, a few years later, after showing the film in front of a large audience

234 | Handbook of Mathematical Science Communication

at Leipzig City Hall, a young man came up to me and informed me that he was also in my film! It turned out to be László Székelyhidi. Cédric had referred to him in his explanations and his name can be seen on the board in the upper left corner.

Looking ahead, I will say that Cédric takes the mission of popularizing mathematics very seriously. A few years later we presented the film together in Seoul, before the opening of an exhibition at the Seoul Museum of Contemporary Art, where *Colors of Math* was shown nonstop in a loop in a huge hall every day for 6 months.

We had to answer questions from Seoul children who lined up in a very long line. Cédric came on stage, took off his shoes as usual and left in only his socks, and then answered every question of the schoolchildren as seriously and in detail.

The next character, Anatoly Fomenko, is known for his mathematical paintings. This is why he was chosen to present mathematics in connection to sight. For many years now he has represented various abstract mathematical objects in his graphic works. So his choice for the role of linking mathematics and sight was logical. In his chapter, Anatoly talked about the Plato problem, which roughly explores how to show the existence of minimal surfaces with a given boundary. As an illustration of minimal surfaces, we took a helical staircase in the form of a helicoid, a spiral part of a meat grinder, a bee honeycomb and also a soap film stretched on a wire against the background of the Moscow skyline as seen from the top of Moscow State University's main building.

Aaditya V. Rangan was responsible for the relationship between mathematics and smell in the film.

Unlike the other chapters, here the relationship was literal — Aaditya is studying the human brain's perception of smell using mathematical methods. Compared to smell, speech is different because in speech not every combination of sounds is possible. He talked about solving the "cocktail party" problem of listening to one specific person in a situation where there are many different conversations going on at the same time. Since not every combination of sounds is a word, speech is structured, and this task is difficult, but solvable.

Aaditya studies what type of brain can be responsible for the sense of smell and how this type of brain can work.

Günter Ziegler represented the perception of mathematics through touch in the film.

We had a long discussion about filming, choosing among his many scientific interests a mathematical story to tell in the film. I visited the Technical University of Berlin, where Günter was working at the time, once a week for a few months. Günter was the only German mathematician in the film, and we wanted the storyline to also show something very Berlin-specific.

As a result, Günter decided to talk about Hilbert's 18th problem, packing space with identical figures. Among the planned locations was the Kaiser Wilhelm Memorial Church. Günter drew attention to the geometric shapes: circles on the floor in front of and inside the church.

In filming this chapter, the cameraman and I tried to see Berlin through the eyes of a mathematician reflecting on the problem and noticing the symmetrical geometric shapes where we normally do not see them.

In his reasoning, Günter easily transitioned to the geometry of spaces larger than three dimensions, such as 4-dimensional space, in which 600 tetrahedrons perfectly cover a three-dimensional sphere.

I was able to organize a very unusual and visually fascinating shoot in a huge, mirrored hall with infinitely reflecting facets. The staff at the Berlin Film Museum with whom I had previously collaborated on an exhibition in Moscow on Marlene Dietrich met us half an hour before the museum opened and made the room available.

We filmed Günter Ziegler just as he was changing jobs and moving with his large team to the Free University. In one of the episodes Günter began to reflect on mathematics next to the still empty bookshelves. He seemed oblivious to the presence of the camera and began to speculate aloud about polyhedrons with more vertices and more faces in spaces of large dimensions.

This went on long enough, but the spectacle of a scientist thinking aloud about his world was absolutely mesmerizing.

I tried to put all the characters in this film into a state where they would forget the presence of the camera and would just start thinking out loud about their tasks. I think it's these episodes that are the most valuable in terms of cinematography.

Jean-Michel Bismut brings together sound and mathematics in the film.

He calls mathematics the music of thinking. In several situations, he talked about trying to mathematically describe the transition from "free" Brownian motion to movement along the shortest possible path, the geodesic path described by classical dynamical systems. Free movement also has an allegorical meaning in this chapter. Jean-Michel talked about the absolute inner freedom that the practice of mathematics gave him.

My film was released in 2012 when there was a surge of civic activity in Russia and thousands of demonstrations. Jean-Michel's words about freedom sounded very relevant and were often quoted in the democratic press. Jean-Michel joked that after the film was released, he would be afraid to travel to Russia because he was afraid of having problems with the authorities.

After making the film, Jean-Michel and I became very good friends. I remember a funny episode after the shooting, we were sitting with Jean-Michel in a cafe in Berlin, and he from time to time called his 90-year-old mother, a passionate music lover, who through the internet watching the ongoing Tchaikovsky competition in Moscow.

With a passion that I used to think only football fans were capable of, she reported on the latest news of the music competition.

The final chapter of the film was about a strange organ of the senses, a sense of balance.

I certainly wanted Maxim Kontsevich in my film. Maxim was the greatest Moscow mathematician of my generation. When I was a student of mathematics at Moscow State University, I even attended a seminar he taught for a while. His seminar was called a seminar for schoolchildren and was considered a continuation of the famous Israel Moiseyevich Gelfand seminar. The last chapter of the film was perhaps the most poetic. Maxim appeared as a searching and doubting scientist.

In the film *Colors of Math*, I have tried not to explain some mathematical subjects so much, as to inspire the viewer to pay attention to it and to the scientists involved in it. This principle reaches its climax in the last chapter. Here Maxim was no longer concerned with making his speech logically coherent and clear, but lets us in very close, we could see his "thoughts out loud".

Maxim says that he tries to unite real numbers and p-adic numbers, to understand how analysis arises in algebra, to understand the mystery of real numbers, and to interpret non-commutative objects through geometric images. Maxim calls this not-yet-existing field noncommutative Keller geometry.

After the release of the film, Maxim told me that there were colleagues who heard his ideas in this seemingly confused speech, and this served to further their dialogue.

After the release of the film and participation in numerous screenings and discussions, I saw that in a sense the target audience was divided — the audience that had nothing to do with mathematics and did not even try to understand it, and at the same time very advanced specialists with an understanding of the field in which Maxim works, were enthusiastic about this chapter. Some of the mathematicians, however, were puzzled by this chapter and I was asked why it was necessary to include in the film, if Maxim "didn't tell me anything" in it.

It is interesting that, thinking out loud, Maxim sometimes uttered words and phrases, which for him meant certain mathematical ideas, but for the audience of the film, they read their own meanings into these phrases.

Here are some examples:

-"What we're doing is building new spaces..."
-"We have to look at everything at the same time, as a single object..."

If one accepts the fact that science is a kind of religion for mathematicians, then the reading out of new meanings by the spectators can probably be compared to the interpretation of some religious texts.

At the same time, there are several ironic episodes in the "equilibrium" chapter of the film. For example, Maxim humorously recounts how he convinces fellow passengers on the train that 8×8 will be 48, sneers at the fact that he has to be responsible for balance in the film, and at the end, after asking if he can jump off the pedestal, he jumps off.

I think that in this way, at the end of the film, I kind of shattered the structure that I seemed to have painstakingly built up over the course of the film — the structure of a film with a perception of mathematics through the senses.

In addition to the cameraman Kostomarov, a young Berlin composer, Mike Schroeder, was also working on this film and he composed original music for each chapter. This music was itself, in my opinion, mathematical, but not in the sense of harmony, but in the sense of being unexpected, like a puzzle.

Colors of Math [7], which was made as a study film, turned out to be incredibly popular and successful, unexpectedly for everyone, including me. It ran in cinemas in St. Petersburg every day for 16 weeks and in Moscow every day for 17 weeks.

At the after-show meetings, I sometimes met people who had seen the film in Siberia and then several times in Moscow. Sometimes they told me that the film was not understandable and they wanted to see it again. The film has been translated into 15 languages.

I still regularly receive letters requesting screenings. And in 2021, 8 years after its release, there are institutes and organizations wanting to organize special screenings with discussions after the film.

The press has written a lot about our film. Here are some quotes:

"New Russian film on math packs cinemas" [1]

"500 visitors flocked... international award-winning film *Colors of Math*" [2]

"Essence of mathematics presented through the five senses — a move that finally works" [5]

"Possibly the best math movie you haven't seen" [4]

"Despise TV — silly snobbery" [5]

"Mathematics poetically filmed" [6]

The Discrete Charm of Geometry, 2015, 67 minutes

In 2013, Alexander Bobenko, director of the Collaborative Research Center "Discretization in Geometry and Dynamics" (CRC DGD) invited me to participate in the Communication and Presentation project and to produce mathematical films.

The first film we made at CRC was called *The Discrete Charm of Geometry* (DCG).

The aim of this project was simple: to dive in and show from the inside how mathematicians work in today's world. Another aim was to try to show how mathematicians work as a team, how they influence each other, and how they exchange ideas.

Unlike the first film, *Colors of Math*, it was possible to make a film using the long observation method, a method in which the director intervened minimally in the development of the situation.

The difficulties of this project were not immediately apparent.

First of all, a spectator's film must always obey the laws of dramaturgy; there is no dramaturgy in real life, you have to see it or invent it.

There was a second difficulty. It seemed that by shooting alongside mathematicians for a long time, you could get very close to the characters. I was very experienced in interviews by that time, but I hadn't realized before — there was a downside to working alongside the characters. When a film-director comes from the outside and has enough experience, he can use the effect of a fellow traveler on a train and encourage his characters to share openly and sincerely. Then there's the question of ethics — how to handle this material.

It is very important to me that in my professional career I have never had any cases where my characters have regretted that they had said too much, that I misused their sincerity. I have always handled the material very carefully and with great respect for the characters. But at work it is customary among colleagues to keep a certain distance, and this is a perfectly healthy state of affairs. However, a few weeks after the various interviews, the characters found that I didn't disappear anywhere, I kept being present, and this influenced subsequent conversations with the camera.

Last but not least, the difficulty was the mathematics itself. How deeply can and should the audience be shown the challenges that the team solved in order to remain in the genre of popular cinema? How ready is the audience to absorb the mathematics and for how long?

There were emotional parts and educational parts to the film. The question of how mathematical ideas can be presented in a popular film probably deserves a separate study.

The structure of the film was defined by an attempt to solve these three aforementioned problems.

The film was shot over the course of a year and a half, and I witnessed the solution of several problems, from their formulation to their publication.

For the film, I decided to choose a few subjects to tell in greater detail and other subjects only partially, "tangentially".

When I started the film, I filmed the mathematical discussions myself. Later, I found partners who were interested in the Russian version of the film, and we were able to involve a wonderful professional camerawoman, Irina Shatalova, for some of the filming, and in some cases Pavel Kostomarov himself filmed the film.

There are several storylines in the film in parallel. In one of them we decided to show the process of reflection, solving a mathematical problem. This storyline, unlike the main fabric of the film, was specially constructed. To do this, we used a problem that one of the characters in the film, Boris Springborn, came up with for a mathematical calendar.

The formulation for mathematicians was: what is the largest triangle in a flat torus? But this problem could also be formulated in terms understandable to younger schoolchildren, using the example of a biscuit that is cut out of dough by shifting the shape parallel upwards and sideways.

I gave the problem to the mathematics professors in the film and asked them to think about it and, if necessary, to make drawings on different glass surfaces. Through these glass surfaces, we filmed close-ups of the characters solving the problem in real time. However, we did not give the audience the actual formulation of the problem until the end of the film, inviting them to enjoy solving it themselves.

Subsequently, I was often invited by various universities and mathematics departments to special screenings, and almost always the students enjoyed solving the problem during the discussion. I must say that there were usually not many people who could solve it during the time the film was being discussed after the film screening and the Q&A session.

The other storyline of the film was a scientific problem about the conformal parametrization of a three-dimensional letter B, i.e., a torus with two holes.

For the mathematicians at the CRC DGD this task was not difficult at all. It was completely solved within a couple of weeks, but because it is geometric and easy to visualize, it proved to be fertile ground for the film. At the end of the film, we even showed how the letter B was printed on a 3D printer at the Technical University of Berlin.

Unlike *Colors of Math*, which portrayed mathematics as the domain of gifted individuals, *The Discrete Charm of Geometry* [8] shows the everyday life of a research team through close observation. The film focuses on scientific communication, collaborative and creative search, intergenerational transfers of knowledge — in other words, on the scientific process itself. There are many characters in the film, but in most of the sequences we see Alexander Bobenko, as head of CRC DGD, linking most of the scientific research and subjects in the film.

One of the episodes deals with the use of the results of the center in architecture. Helmut Pottmann showed examples of how modern mathematics was used in the construction of the roof of the Islamic Hall in the Louvre and in the design of the new pavilions at the Eiffel Tower.

In this episode, the audience was no longer questioning why the mathematics was needed, and the entertainment part was already there. This gave me the opportunity to engage the audience a little deeper into the subject of mathematics itself. We used this episode to include an educational part, namely the definition of principal curvature parameterization and conical mesh.

During the filming period the famous physicist Freeman Dyson visited Berlin. One of the main characters in the film, Alexander Its, had worked for many years on proving Dyson's hypotheses and was very enthusiastic about the opportunity to meet and talk to him.

Thus the episode of Its and Bobenko meeting and talking to Dyson appeared in our film. Surprisingly, Dyson was hardly ever filmed. He revealed that Stanley Kubrick shot him in his famous film *2001: A Space Odyssey*, but in the final cut an episode involving him was never included in the film. And Dyson was also filmed by Japanese documentary filmmakers, but for some reason in a film about butterflies. I am glad that we were able to show in the film my characters' interaction with Freeman Dyson, already a classic in his lifetime. Its and Bobenko talked to him at Berlin's Tiergarten café about the philosophy of the CRC DGD and, in particular, about the discrete ellipsoid problem.

In the mathematical parts of the film Emanuel Huhnen-Venedey talks about Supercyclidic nets, Tim Hoffman and Wolfgang Schief about the problem of the "correct" definition of discrete curvature, Yuri Suris about constructing discrete elliptic coordinates, Ulrich Pinkall about using quaternions to describe motion in space of surfaces and, in connection with this, he demonstrates what "simply-connected group" means.

At the end of the film, in the final credits, we have a list of 10 papers which have been written during the making of the film and which are mentioned in the film.

At one point, Alexander Bobenko told us about his "dream problem" and even formulated it while on a boat in a lake. Amazingly this difficult classical problem has been solved after 5 years by him and his colleagues at CRC DGD, and in this way, we have in our hands a unique material which we are going to use in the next new film.

However, mathematical plots have been embedded in the usual purely cinematic fabric of the film. There is a poetic line in the film — the lake, the water, the boat. We don't immediately show who is in the boat.

And the most emotional part of the film is followed by an episode that simply captures the wind in the corridors of the university. Here, the sound solution of the episode is important. Alexander Sokurov, one of the most highly respected Russian directors, praised the sound in the film, on which Arthur Khairullin worked. Sometimes montage transitions were made through rhythmic soundscapes — for example the sound of fingers tapping on the cup in the audience was transformed into the sound of steps in the city in the next episode, or a scientific dispute at the blackboard and the following episode with a knife flying at a target, or simply the ticking of the clock, which increased the tension.

A large part of the film is observation of the characters' daily lives, but each time through the prism of their mathematical mindset.

We showed an absent minded young scientist, Stefan Sechelmann, distracted by his thoughts, counting during dance lessons, practicing throwing a blade at a target, improvising on the piano, constructing a new musical instrument, and even thinking about mathematics while riding a bicycle. In one episode, Professor Suris literally has his daughter jumping on his head while he talks about mathematics.

Emanuel Huhnen-Venedey shares doubts about whether he can hold his own in the academic world, even with a very successful doctoral thesis. In his spare time, he takes up rock climbing, but the episode is edited to show taking up mathematics is riskier and more significant for him than rock climbing.

The audience of the film, sometimes far removed from mathematics, could nevertheless follow the non-mathematical narration of the film, and feel a sense of involvement with the action, as the mathematical parts were not too long.

In this artistic dimension of the film the critical point, the climax, was the question of what the characters were prepared to sacrifice to achieve success in their science. This question is a tough one, because a profession, a job, is a big part of life. Even though the answers mostly consisted of our heroes themselves regretting that they were not prepared to sacrifice too much, the question itself and the reactions to the question were a strong emotional component of the film.

The legendary mathematician John Nash also appears briefly in this episode and this is one of the last shots of his life. The story of his appearance in our film is as follows.

The Abel Prize Board commissioned me to make short film portraits of the laureates in order to show them at the prize-giving ceremony. In 2015, the prize was awarded to Louis Nirenberg and John Nash. We were filming Nash at Princeton, and I told him that I had arranged to meet Freeman Dyson after the shoot and show him the material we had shot with him in Berlin and were going to use in the film. And Nash then offered to take us to his old friend Dyson himself. So cameraman Pavel Kostomarov and I witnessed the last meeting and conversation of these great scientists.

They remembered Albert Einstein, talked about John von Neumann and Andre Weil. I decided to include only a very brief episode from this meeting in our DCG film, so as not to shift the focus from the main story about the work of CRC DGD. As you know, John Nash was tragically killed in a car accident on his way back from Oslo after receiving the Abel Prize, and Freeman Dyson also died in 2020.

I think the material shot on that day is of great value to the history of science and is certainly waiting to be used in future films.

The Discrete Charm of Geometry had many subjects and film critics have compared the structure of the film to an impressionist painting, where individual strokes form an overall picture.

This film was also often shown at all kinds of conferences, at special shows in universities and schools all over the world, and was running in Russian cinemas. There was a lot of press coverage of the film, for example here:

- "...If you are to watch one movie this year, perhaps this should be it..." — Robert Harington about *The Discrete Charm of Geometry* for the Scholarly Kitchen [9].
- On "The Discrete Charm of Geometry", https://3quarksdaily.com/ 3quarksdaily/2016/09/on-the-discrete-charm-of-geometry.html [10].
- Discretization in Geometry and Dynamics — press release, https://www. discretization.de/news/2015/08/21/discrete-charm-geometry/ [11].

I remember a funny short review someone wrote on Facebook like this:

"...mathematicians who have retreated from science should watch carefully. You might get black with envy."

It was also a pleasure to hear from an audience member that, after our film, he had decided to return to the abandoned dissertation.

The Discrete Charm of Geometry begins with a mathematical problem and ends with a formulation of a problem for the audience.

One of the film critics wrote that while the mathematicians were solving their problems, I was solving mine — no easier than the one solved by the characters in the film. Namely, I was trying to find the limits — how much mathematics you can show on the screen, how much detail you can go into the closed areas. Whether those boundaries have been found is an open question.

After *The Discrete Charm of Geometry* I have produced other mathematical films, including:

- *Georg Cantor, the Discoverer of Infinity*, 2018, 45 minutes.
- *Head, Heart and Soul*, 2019, 54 minutes.
- *Math Circles Around the World*, 2020, 20 minutes and 45 minutes.
- *Lobachevsky Space*, 2021, 75 minutes.

In conclusion, I think that the question of how deep and how one can show mathematics in cinematography requires further investigation!

Acknowledgments

The author is supported by the DFG Collaborative Research Center TRR 109 "Discretization in Geometry and Dynamics".

References

[1] The Moscow News. (May 2013) New Russian film on math packs cinemas. https://archive.ph/kcrwT

[2] Discretization in Geometry and Dynamics. (May 2015). Colors of Math. https://www.discretization.de/media/filer_public/2014/05/16/colors_of_math_eremenko.pdf.

[3] Moskauer Deutsche Zeitung. (July 2012) 5 Sinne + 1 Regisseurin. https://old.mdz-moskau.eu/5sinne1regisseurin/

[4] Karlsruhe Institute of Technology. (June 2013). Colors of Math. https://topology.math.kit.edu/english/347.php

[5] Kovrigina, A. (April 2013). "Despise TV — silly snobbery". *Lcd Televisions*. http://lcdtelevisionse.blogspot.com/2013/04/ekaterina-eremenko-despise-tv-silly.html

[6] Freie Univerisität Berlin. (June 2012). Mathematik poetisch verfilmt. https://www.fu-berlin.de/campusleben/forschen/2012/120611_mathefilm/

[7] Eremenko, E. (February 2013). *Colors of Math*. https://www.imdb.com/title/tt2293548/.

[8] Eremenko, E. (October 2015). *The Discrete Charm of Geometry*. https://www.imdb.com/title/tt4885016/?ref_=fn_al_tt_1.

[9] Harington, R. (July 2016). The Discrete Charm of Geometry — A review. *The Scholarly Kitchen*. https://scholarlykitchen.sspnet.org/2016/07/27/the-discrete-charm-of-geometry-a-review/.

[10] Pierer, C. (September 2016). On "The Discrete Charm of Geometry". *3 Quarks Daily*. https://3quarksdaily.com/3quarksdaily/2016/09/on-the-discrete-charm-of-geometry.html.

[11] The Discrete Charm of Geometry. (August 2015). *Discretization in Geometry and Dynamics SFB Transregio 109*. https://www.discretization.de/news/2015/08/21/discrete-charm-geometry/

https://doi.org/10.1142/9789811253072_0014

Chapter 13

The Project Mathematics and Culture: Films on Art and Mathematics

Michele Emmer

Math Department, Università Sapienza, Rome, Italy;
Istituto Veneto Scienze, Lettere ed Arti, Venice, Italy
michele.emmer@uniroma1.it

Prologue

Sir Roger Penrose, at the age of 90 (he was born on August 8, 1931) was awarded the 2020 Nobel Prize in Physics for his research on the black holes of the universe, which he started many decades ago. Penrose has always been an interesting and creative person (Figure 1). Among his many interests are Escher's works and quasi crystals. I had discovered Escher's works in the early sixties. A few years later I started thinking about making films on the relationship between mathematics and art. I started writing articles and books on the subject, and over the years I thought about realizing not only films but also exhibitions, conferences, and books collaborating with mathematicians and artists from different parts of the world. I remembered the book I saw on Escher's work [1], and reading it I noticed that the Dutch artist collaborated with two famous mathematicians, Penrose and Donald Coxeter (Figure 2). I contacted both of them to collaborate in the making of a film on Escher [2], and a few years later in the organization of the first international conference on the artist's work [3] and a major exhibition to be inaugurated in Rome by the Queen of Holland in 1985 [4]. My family is of (centuries ago) Dutch origin. In the meantime, I had started working as a mathematician on minimal surfaces and surfaces with assigned curvature. In short, soap bubbles. I consider making a film on soap bubbles with two US mathematicians I knew well using paintings on soap bubbles. I discovered a whole fascinating world

Figure 1: Sir Roger Penrose, frame from the film *Escher* © M. Emmer.

Figure 2: H.S.M. Coxeter, frame from the film *Escher* © M. Emmer.

of two parallel worlds, art and science, which sometimes communicate between them [5].

The Idea of the Film Project

From the very beginning of the project there was the idea of focusing on the influence and the connections of mathematics and culture, using the most important visual tool: film. The idea was to show a single theme, such as soap bubbles or topology, or the works of Escher, which originated the interests of artists and mathematicians and to consider if it was possible to exchange ideas between scientists and artists — all these themes, however, following the teachings of my father who was a filmmaker, were to be expressed using images alone. I had to build a sequence of images that was first of all interesting from the point of view of cinema. A film is not an essay or a picture book. The visual elements fascinate *per se*. So the idea is to use few words and many images, using all the possibilities of the language of cinema. The theoretical study on the links between mathematics and art has been amply treated in the books I have written and the volumes of the proceedings of the International Congress on *Mathematics and Culture* in Venice [6]. As these were the general lines of the project, it was quite natural to consider as part of it the organization of exhibitions (many were made in the next years), congresses and seminars, and the publishing of books (with many illustrations!), including the two volumes of *The Visual Mind: Art and Mathematics* in the Leonardo Book Series (MIT Press) [7,8] and even theses for students in mathematics, history of art, and architecture. Today, more than 45 years later, it is easy to say that the project went far beyond expectations.

In 1970, I started my professional career at the university and one of the most difficult things to do in an Italian university is to be involved in a field connecting two or more different areas. During the last 45 years, I have been invited to many Italian and foreign universities to show and discuss my movies. But when I first showed one of my movies in Rome in 1981 to a wide public audience, mathematicians of my department told me that it was not good for the reputation of our department. Mathematicians just demonstrate theorems! This is the main reason why almost all my movies have been made abroad — in Europe, US, Canada, Japan, and even India. And the same is true for the publication of books and proceedings of congresses organized abroad or with the help of non-Italian mathematicians. This is the reason why it was possible to organize the *Mathematics and Culture* congress in Venice, starting only in 1997 (Figure 3). I can say I was the first person to promote this idea of mathematics and culture in Italy.

Figure 3: M. Paladino, *Without title*, Poster *Imagine math 7*, Venice, 2019, private property.

Did I have in mind to make educational films and to popularize mathematics? I wanted to make docufilms by letting mathematicians and artists talk, but above all by showing meaningful images for both mathematics and art, making completely professional films. Most of the films were seen and used around the world in different languages in universities, museums, art exhibitions, schools, televisions, even at the *Venice Art Biennale* [9,10]. At least one film is a true story, with characters, all in animation, based on *Flatland*, the book by Edwin A. Abbott [11], with original music by Ennio Morricone. The first exhibition I organized was the one on Escher in 1985, and the latest in 2019 on soap bubbles in one of the great museums of the Italian Renaissance, the *National Gallery of Umbria* in Perugia [12].

Why would a mathematician who worked on minimal surfaces take up making films?

My father Luciano Emmer was a famous Italian filmmaker. He made not only movies but many documentaries on art, for example a documentary with Picasso in 1954 [13] and on Leonardo da Vinci [14] that won a *Silver Lion* at the Venice Film Festival in 1948. Marcello Mastroianni made his first film with him, *Domenica d'agosto* in 1949 [15]. I learned

Figure 4: The author in the film *Camilla*, 1954, by Luciano Emmer. © M. Emmer.

a lot from my father, and I found myself a capable filmmaker. So for me thinking of making a film became quite natural. When I was young I was always involved in filmmaking — as collaborator, organizer, and even actor — in several of my father's movies [16] (Figure 4).

The Math and Culture Project

The project started in 1976, together with the idea of the film series *Art and Mathematics*, produced by my father and the Italian State television. Or better, that is the year when I started thinking of the project because of essentially three reasons. The first was of course a film on the works of M.C. Escher with Roger Penrose and Donald Coxeter. From the first time I saw Escher's engravings my purpose was to make a film only on him.

The Fantastic World of M.C. Escher, the Movie

My idea was to use the technique of animation (no computer graphics of course since the drawings for animation were supposed to be *a la mode d' Escher*) in order to make his works really three-dimensional using time, movement, and music. Animation was made frame by frame using drawings and photos of the original works of the Dutch artist. To make the animation you must cut

Figure 5:   Animation frame, Escher's movie © M. Emmer.

the artwork in pieces and this is of course impossible with the original one. Something that Escher himself suggested; he personally was involved in a short film with several animations of his works before his death in 1972. The first conference I organized on Escher was held in Rome in 1985 [3], organized not only with Coxeter and Penrose but also with Marianne Teuber, and a second one in 1998, in Rome again, with Doris Schattschneider [17–19]. The interview with Penrose and Coxeter were filmed in Rome, with Bruno Ernst in Utrecht and Caroline Macgillavry in Amsterdam.

My idea was to film all over the world where the artists and mathematicians involved in the film were working. My father was the producer for all of them, so I was able to find more support. In any case the Escher film project was postponed. I only filmed three works of Escher and inserted them in the film *Möbius Band.* It would take me more than 10 years to complete the film on Escher in 1990 (Figure 5).

The financial support of RAI (State Italian Television) was not sufficient to make my first four movies; it was out of the question to make a film on Escher using animations, as this technique is very expensive. Financial support from RAI was enough just to make a film, entirely in a studio, with a person talking all the time; their idea of an educational TV series. The main reason why for the first time RAI decided to make films on math was that my father was the producer. They trusted him.

Minimal Surfaces: Soap Bubbles

The second reason: in 1976, I was at the University of Trento in northern Italy. I was working in the area called the *Calculus of Variations*, in particular, *Minimal Surfaces* and *Capillarity Problems* [20]. I was very lucky to start working with Mario Miranda, who was working with the famous mathematician Ennio De Giorgi. I also met Enrico Giusti and Enrico Bombieri. It was the period in which in the investigations of *Partial Differential Equations*, the *Calculus of Variations* and the *Perimeter Theory* — first introduced by Renato Caccioppoli in the fifties and then developed by De Giorgi and Miranda [21–24] — were ongoing at the Scuola Normale Superiore of Pisa, one of the best universities in the world. In 1974, Enrico Bombieri received the Fields medal, the most important prize in math, in the absence of the Nobel prize for mathematics. By chance, I was in the right place at the right time.

In Trento I met Fred Almgren and Jean Taylor, Jean proved a famous result on singularities of the edges that soap films generate when they meet [25]. Again in 1976, the journal *Scientific American* asked Jean Taylor and Fred Almgren (they had married a few months before) to write a paper on the more recent results on the topic of *Minimal Surfaces and Soap Bubbles* [26]. When Almgren and Taylor came to Trento, the issue in the *Scientific American* had just been published. The pictures of the article and the cover were really beautiful and interesting. I do not remember why, but when I looked at the pictures I had the idea of making a film on soap film to show its shapes and geometry in the greatest possible detail, using the rallenti technique. Both Fred Almgren and Jean Taylor were very interested in my project. My idea was not to make a *small* scientific film, like a sort of scientific commercial just to show some small experiments with soap bubbles and soap films. I was attracted by the phenomena of soap films because they were visually interesting and I thought that the technique of filming them would increase general interest and fascination about them (Figure 6).

In 1982, I was invited by Algrem to Princeton University. I showed the preliminary version of the film in his course on the *Calculus of Variations*. I was also invited by Enrico Bombieri to present the movie at the Institute of Advanced Study in Princeton. Near the Math Department was the

Figure 6: M. Emmer, E. Bisignani, *Soapy Hypercube*, photo, Biennale Arte Venezia, 1986.

Figure 7: A. Romako (copy), *Two Boys Playing with Soap Bubbles*, end XIX century, oil on canvas, private property.

Art Department and I started looking for paintings and engravings on soap bubbles in the last centuries. I discovered a real universe! I have written two volumes on soap bubbles in art and science and I am still working on this theme, for example, see the catalogue of the exhibition in Perugia [12] (Figure 7).

The *Möbius Band*

Now the third reason for starting the *Art and Mathematics* project. I was working at the University of Trento while my family lived in Rome. Every Friday, I left Trento to go to Rome and then on Monday, I travelled back to Trento. I have always been a lover of art, of all kinds, of any culture and period. I read in a newspaper of an exhibition in Parma that was dedicated to one of the most important artists of last century: Max Bill. I already knew some of the sculptures of the Swiss artist but I had not visited a large exhibition like the one in Parma before. I decided to stop on my way back to Rome to see the exhibition [27]. The topological sculptures of Bill were a real discovery for me. Years before, I had seen a large exhibition of the works of Henry Moore in Florence and of many other artists, but Bill's works almost immediately gave me the impression of *Visual Mathematics* [28]. The *Endless Ribbon* — an enormous Möbius band made of granite — was a real revelation. Its shape, its physical nature, and its tridimensional reality made it live in space. It was a mathematical form — alive. This was the missing idea: mathematicians throughout all the historical periods and in all civilizations have created shapes, forms, and relationships. Some of these shapes and relationships are really visual; they can be made visible. This idea is behind the great success of using computer graphics in some sectors of mathematics. The project was becoming clearer: to make films, in which to compare the same theme from a mathematical and artistic point of view, asking for the opinion of mathematicians and artists. To *make visible the invisible* like the artist David Brisson says in the 1984 film *Dimensions* with Thomas Banchoff. The themes of the first two films were: soap bubbles and topology, in particular the Möbius band. To have more visual ideas and objects to film, I finally decided to include in all the movies the connections between mathematics and architecture, all the other sciences, in particular biology and physics, not excluding literature and even poetry as a poem of Charles Olson on the *Möbius Band*.

I contacted Max Bill by writing him a letter (Figure 8). He was very kind; he invited me and my troupe to his house in Zurich, and he gave me permission to film everything I was interested in, including his fabulous collection of contemporary art. There was one exception. It was

Figure 8:   Max Bill, frame from the film *Möbius Band* © M. Emmer.

strictly forbidden to film a little window in which he had his collection of topological forms made of paper — very small objects — which was his data for future works. He was afraid that someone could see his projects and copy them. We then became friends and made two exhibitions and another film on *Ars Combinatoria* together [29]. Bill said in the movie: "At that time, I was looking for the basis of the development of figures, and I went back to music knowing that it is based on the laws of mathematics, and that it is possible to find rules behind the organization of a theme. This was the beginning of the idea of variations. I then tried some types of variations but in a limited system. In other works, I tried to push a theme to its limit." We both were on the editorial board of the journal *Leonardo*, at that time published by Pergamon Press. For my book *The Visual Mind: Art and Mathematics* [30], Bill rewrote the title and made some changes to his famous paper originally written in 1949 on a mathematical approach to art. Two of Bill's works are reproduced on the front and back cover of the book. The volume *The Visual Mind 2* is dedicated to Max Bill [31]. In my annual congress in Venice, I dedicated a special session to Max Bill in 2014, 20 years after his death [32]. In my paper [33], I show pictures of the making of a large sculpture by Max Bill, starting from the choice of the marble in Carrara and the transportation to Germany [34]. A large exhibition of Bill's work was organized in the *Palazzo Reale* in Milan in

2008 [35]. There was a part dedicated to his topological sculptures, not including the largest and the most famous *Endless Ribbon*. It was his dream to have a large exhibition of his topological works.

Other Few Examples

Labyrinths

Mathematician Anthony Phillips described the mathematical structures of labyrinths like the Cretan Maze and its connections with music. It is possible to connect all Cretan-type labyrinths to music, in such a way that it is possible to recognize each type using music (Figure 9).

The musician David Macbride plays labyrinthic music in the film. The Italian painter Fabrizio Clerici produced a very large painting for the film and for the exhibition *The Eye of the Horus: Mathematics and the Imagination* [36,37] that was held in Italy in 1989. I made a film for the Italian public television, RAI, in the series *The Great Exhibitions of the Year* for the exhibition. These are words of the artist Fabrizio Clerici on the theme of the labyrinth: "I think that for an artist who was first educated as an architect and wound up as a painter, the labyrinth is a natural step: it's the kind of architecture a painter is continually rebuilding in various ways. I think I drew my first labyrinth when I was a child, unthinkingly, without knowing

Figure 9:   Cretan labyrinth, frame from the movie © M. Emmer.

it was a labyrinth; like a geometrical design. But an actual labyrinth, a kind of trap, that happened much later, around 1945, just after the war. Since the edifice was a labyrinth, the personage that must dominate the entire situation had to be the Minotaur who is accusing the mother which bore this monster." The final part is filmed by a helicopter flying over the XVII century bush labyrinth in the gardens of Villa Pisani near Padua in Italy.

Dimensions

This film was made with the mathematician Thomas Banchoff and art historian Linda D. Henderson. Banchoff realized the first animation of a hypercube, while Henderson wrote in 1983 the seminal book *The Four Dimension and Non-Euclidean Geometry in Modern Art* (Princeton University Press) [38]. The idea was to show the four-dimensional polytopes using the different projections in three-dimensional space using time and movement to make the viewers feel the higher dimensions. The two American artists Harriet and David Brisson were involved in the film. David: "The reason why I am interested in visualization of the materials is that a lot of it is hard to understand in a verbal sense. But when they were made into visual images then it was possible for me to understand very complicated ideas. And I felt that there were a lot of other people who could profit by making or by having the experience of these ideas expressed to them visually when they were not ready to understand these ideas in purely verbal or mathematical terms. A lot of these mathematical ideas are very interesting and beautiful when converted to 3-dimensional form."

Cinema and Mathematics

I have been always interested in cinema and, in particular, the connections between cinema and mathematics. I wrote in 2017: [39, see also 40–42] "Many years have passed since James Stewart played a poet and teacher of English literature at a university in the USA in a comedy film. He was in constant conflict with the scientists of his university and considered scientific culture and mathematics in particular arid and un-educational. One day his son's elementary school teacher discovers that he is a small mathematical genius. Or rather a boy with an outstanding ability to perform mental calculations. Of course, this is not what mathematicians do, but it comes quite naturally in everyday language and thus in films, to call mathematical genius a kid who does very fast calculations. The teacher accidentally discovers this ability and happily goes to visit his parents. She explains that the child is a prodigy of mathematics, and the father Stewart with a paling face, puts a hand on his wife's shoulder to comfort her. After the teacher leaves, he begs his son not to tell anyone of his ability, the source of so much possible trouble, especially because passers-by in the street could comment: "He is a mathematician!" Stewart pronounces the sentence in disgust, commenting "We would never want something like that to happen." The real star of the film was Brigitte Bardot, in the role as herself. So much so that the movie was called *Dear Brigitte* [43]. She was the object of the dreams of Stewart's son in the movie, much more than numbers. As years passed by, the attitude toward mathematicians in film radically changed. It even became possible to win an Oscar with films about mathematicians! I have written several papers and books on mathematics and cinema, including a book in Italian in 2017. I organized the World Mathematical Year 2000, a festival of math and cinema at the University of Bologna with thousands of spectators for 3 months. Filmmakers, actors, and writers participated in the many events. Proceedings were published by Springer in 2005 [44]. One of the films was the well-known *Fermat's Last Theorem* [45] by filmmaker Simon Singh. Simon came three times to my conference in Venice, writing papers on his experience as writer and filmmaker in the proceedings [46,47] (Figure 10).

The Audience

As these were the general outlines of the project, it was quite natural to consider as part of it the organization of exhibitions (many were presented in subsequent years), congresses, and completely new courses, including teaching for 3 years at the *International School of Scientific Journalism* at the SISSA in Trieste, creating a completely new course on *Space and Form*, for final year students of math, and hosting seminars and publishing books (with many illustrations) in industrial design for 10 years. The last book

Figure 10:   S. Singh, frame from the film *Math & Culture*, M. Emmer, Venice, 2008 © M. Emmer.

Imagine Math 7 by Springer Nature was published in October 2020 [48], which detailed the proceedings of the March 2019 conference in Venice. After 2 years, new volume, *Imagine Math 8* was published in 2022 [49]. Of course it was not possible to organize anything in Italy in 2020 and also in spring 2021. So the new volume is not about the proceedings of a Venice conference. I wrote a paper on Venice at the time of the virus for a special book edited by Springer, *Math in the Time Corona* [50]. In 2019, there were two last international exhibitions, one in Venice for famous Italian contemporary artist, Mimmo Paladino, who realized ten original posters for the Venice conference, one of them being the cover of *Imagine Math 7*. The second one, already mentioned previously, is in Perugia and looks at soap bubbles in art and science. It had more than 40,000 visitors [12]. So the project was intended for really a large audience. Hundred of thousands of people have been involved: visiting the exhibitions, taking part in the congresses, studying in courses, and using the books by MIT Press and Springer as references. Films have been translated in French, English, and Spanish, and diffused all over the world.

Conclusions

A film is not the best tool for explaning or learning. A film can, in a short amount of time, provide ideas and suggestions, and create stimuli and emotions. A film can generate interest, even enthusiasm. Looking at an engaging, pleasant film can stimulate the audience to learn more, both in the artistic and the mathematical fields. In this sense I consider my films educational, but only with this meaning.

The fact that the films are not just educational, on the contrary, is the secret of their success, as with the movie *Soap Bubbles*, even after 40 years. The most beautiful sequences I have ever made — soap films dancing with the music *Invitation to the Dance (Le Spectre de la Rose)* by Carl Maria von Weber — were included in the *Video Math Festival* selection for the International Congress of Mathematicians in Berlin in 1998 and in the *European Congress of Mathematics* in Barcelona in 2000 [50]. The movie is still circulating. It was screened at the entrance of the exhibition in Perugia in 2019. Both the films, of the exhibition and of a 2 minutes sequence of the film on soap bubbles, can be seen at the website of the Galleria Nazionale in Perugia [51]. In 2009, I wrote a 400-page book on soap bubbles in art and science that received one of the most important awards for an Italian essay, the *Premio Letterario Viareggio* [5]. Ennio Morricone made the original music of the battle for my animated film *Flatland* with real objects, and the film was invited to several film festivals and studied by films critics. The same for the Escher movies with many animations in 2D. I also gave courses on how to make 2D animation with real objects and drawings (Figure 11).

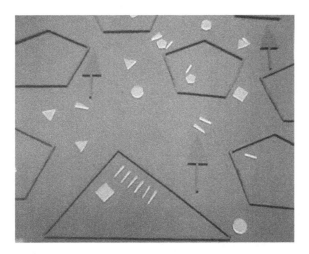

Figure 11: Frame from the film *Flatland* by M. Emmer © M. Emmer.

I organized the first large and traveling exhibition in Italy on art and math, *The Eye of Horus,* in Bologna, Milan, Parma, and Rome in 1989 [35,36]. All of my films were part of the exhibitions.

The *Art and Mathematics* Film Project

Art and Mathematics: my films, produced by Film 7 International and L. Emmer prod., Rome (1979–2013). All films are in color, (most are) 27 minutes and available in DVD in various languages.

Möbius Band (1979)

Soap Bubbles (1979)

Platonic Solids (1979)

Symmetry and Tessellations (1979)

Dimensions (1982)

M.C. Escher: Symmetry and Space (1982)

Spirals (1982)

Helices (1982)

Ars Combinatoria (1984)

M.C. Escher: Geometries and Impossible Worlds (1984)

Knots (1984)

Geometry (1984)

Labyrinths (1987)

Computers (1987)

Avventura del quadrato (1987)

Figure geometriche (1987)

L'occhio di Horus, 35 min, RAI prod. (1989)

Metamorfosi di Fabrizio Clerici, 7 min, animated (1991)

Venezia perfetta, 20 min (1993)

The Fantastic World of M.C. Escher, 50 min, produced by Emmer and distributed in the US by Acorn (1998)

Flatland, 23 min, animated, music by Ennio Morricone (1999)

F. Armati and M. Emmer, eds., *Ricordando Fabrizio Clerici*, AICS prod. (1994)

Matematici part 1, part 2, 60 min, Città della Scienza, Naples (1996)

Ennio De Giorgi, video interview, 75 min, UMI (1997)

Venice 2008, Emmer and Villarreal prod., Rome (2008)

Il Leon musico fa le bolle, 35 min, La Biennale di Venezia (2013)

References

[1] Escher, M. C. (1967). *The Graphic Work of M.C. Escher*. London: MacDonad.

[2] Emmer, M. (1998). *The Fantastic World of M.C. Escher*. Rome: RAI, Film 7 International, Emmer Productions.

[3] Coxeter, H. S. M., Emmer, M., Penrose, R., and Teuber, M. (eds.) (1986). *M.C. Escher, Art and Science*. North-Holland, Amsterdam: Elsevier.

[4] Emmer, M. and van Vlanderen, C. (eds.) (1985). *Maurits C. Escher*, catalogue of the exhibition, March 28 — May 12, 1985, Istituto Olandese di Roma, Rome with papers by J. Offerhaus, G. Escher, H. S. M. Coxeter, C. H. MacGillavry, M. Emmer, exhibition curator Kees Broos.

[5] Emmer, M. (2009). *Bolle di sapone tra matematica e arte*. Bollati Boringhieri, Torino. The Best Italian Essay, Premio Letterario Viareggio 2010.

[6] Emmer, M. (ed.) *Mathematics and Culture, series of Proceeding of the International Congress in Venice* (1997–2011) (Vol. 20). Milano, Berlin: Springer verlag, in Italian and English; *Imagine Math* (2011–2020). New series of the Proceedings (Vol. 8), Springer verlag.

[7] Emmer, M. (ed.) (1993). *The Visual Mind: Art and Mathematics*. Cambridge, MA: MIT Press.

[8] Emmer, M. (ed.) (2005). *The Visual Mind II: Art and Mathematics*. Cambridge, MA: MIT Press.

[9] Emmer, M. (1986). "Excerpt from the film series *Art and Mathematics*". In R. Ascott (ed.), *Arte e Scienza: Tecnologia e Informatica*, Catalogue of the exhibition, Biennale d'Arte di Venezia (p. 83). Venice: Biennale Venezia Publ.

[10] Emmer, M. (1986). "Lo spazio tra matematica ed arte". In G. Macchi (ed.), *Arte e Scienza: Spazio*. Catalogue of the exhibition, Biennale d'Arte di Venezia (pp. 36–39). Venice: Biennale Venezia Publ., Scenes from the film *Möbius Band* (p. 72).

[11] Emmer, M. (ed.) (2008). *Flatland by E. A. Abbott*, English & Italian version, Bollati Boringhieri, Torino including DVD, Film 7 International & Emmer prod., Rome.

[12] Emmer, M. and Pierini, M. (eds.) (2019). *Soap Bubbles: Forms of Utopia Between Vanitas, Art and Science.* Milano: Silvana Publ.

[13] Emmer, L. (1954). *Picasso,* with Pablo Picasso, script by L. Emmer and S. Amidei. Rome: Rizzoli Production.

[14] Emmer, M. (2009). *A Film on Leonardo da Vinci by Luciano Emmer,* Leonardo (Vol. 42, pp. 449–453).

[15] Emmer, L. (1950). *Domenica d'Agosto,* story by S. Amidei, screenplay by F. Brusati, L. Emmer, G. Macchi, and C. Zavattini. Rome: Colonna Film Prod.

[16] Emmer, L. (1954). *Camilla,* story and screenplay by L. Emmer, E. Flaiano, and R. Sonego, Video prod. Paris: Rome and Cormoran prod.

[17] Emmer, M. and Schattshneider, D. (eds.) (1998). *M.C. Escher's Legacy.* Berlin: Springer verlag. Proceedings of the conference in Rome and Ravello.

[18] Emmer, M. (ed.) (1998). *1898–1998 Escher.* Catalogue of the exhibition in Rome. Roma & Ravello: Diagonale publ.

[19] Barucci, V. (ed.) (1998). *Homage to Escher,* catalogue of the exhibition in Rome, with paper by M. Calvesi, M. Emmer, and D. Schattschneider, based on an idea of M. Emmer & D. Schattschneider. Rome: Diagonale publ.

[20] Finn, R. (1986). *Equilibrium Capillary Surfaces.* See the Emmer's results, Berlin: Springer.

[21] De Giorgi, E., Colombini, F., and Piccinini, L. C. (1972). *Frontiere orientate di misura minima e questioni collegate,* Scuola Normale Superiore Pisa, Classe di Scienze, Pisa.

[22] Miranda, M. (2006). *Superfici Minime e problema di Plateau,* quaderni Università di Lecce, 1/2006.

[23] Giusti, E. (1984). *Minimal Surfaces and Functions of Bounded Variation,* Birkhäuser, Basel.

[24] Bombieri, E. (ed.) (2010). *Geometric Measure Theory and Minimal Surfaces,* CIME (Centro Internazionale Matematico estivo), Varenna, August 24–September 2, 1972, CIME Summer Schools, n. 61. Berlin: Springer verlag. Reprinted 1st edn., Cremonese, Rome, 1972. Papers by W. K. Allard, F. J. Algrem Jr., E. Giusti, J. Guckenheimer, J. Taylor, M. Miranda and L.C. Piccinini.

[25] Taylor, J. E. (1976). "The structure of singularities in soap-bubble-like and soap-film-like minimal surfaces". *Annals of Mathematics,* Second Series, 103(3), 489–539.

[26] Almgren, F. and Taylor, J. (1976). "The geometry of soap bubbles and soap films". *Scientific American*, July 1.

[27] Quintavalle, A. (ed.) (1977). *MAX BILL*, Università Comune Provincia di Parma, Italy.

[28] Emmer, M. (1980). "Visual art and mathematics: The *Möbius Band*". *Leonardo* 13(2), 108–111.

[29] Emmer, M. (1984). *Ars Combinatoria*, 25 m., color, DVD, RAI & Film 7 International, Rome: Emmer prod.

[30] Bill, M. (1949). *The Mathematical Way of Thinking in the Visual Art of Our Time*, originally published in German in Werk, n. 3. Winterthur; reprinted by the author with changes by the title *The Mathematical Approach in Contemporary Art* in Emmer, M. (ed.) (1993). *The Visual Mind: Art and Mathematics* (pp. 5–9). Cambridge, MA: MIT Press.

[31] Emmer, M. (ed.) (2005). *The Visual Mind II: Art and Mathematics*. Cambridge, MA: MIT Press.

[32] Emmer, M., Abate, M., and Villarreal, M. (eds.) (2015). *Imagine Maths 4: between Culture and Mathematics* (pp. 5–44). Bologna: UMI & IVSLA.

[33] Emmer, M. *Max Bill: A Journey through Memories*, in [29] (pp. 29–41).

[34] Spies, W. (1986). *Kontinuität, Granit-Monolith von Max Bill*. Deutsche Bank, Dortmund.

[35] Buchsteiner, T. and Letze, O. (eds.) (2006). *Bill*. Milano: Electa.

[36] Emmer, M. (ed.) (1989). *L'occhio di Horus. Itinerari nell'immaginario matematico*, Bologna, Parma, Milano, Roma. Catalogue of the exhibition, Istituto della Enciclopedia Italiana, Roma.

[37] Emmer, M. (1989). Script & Final editing, *L'occhio di Horus*, RAI Italian State Television, series *the Most Important Exhibitions of the Year*, 35 m.

[38] Henderson, L. D. (1983). *The Fourth Dimension, Non-Euclidean Geometry and Modern Art*. Princeton: Princeton University Press; new and revised edition with a new introduction, (2013). Boston: MIT Press.

[39] Emmer, M. (2013). *Numeri immaginari: matematica e cinema*. Torino: Bollati Boringhieri.

[40] Emmer, M. (2006). *Visibili armonie: arte, cinema, teatro e matematica*. Torino: Bollati Boringhieri.

[41] Emmer, M. (2016). "Cinema, Literature and Mathematics: Recent Developments". In M. Emmer *et al.* (eds.), *Imagine Math 5* (pp. 75–90), IVSLA & UMI, Bologna: Unione Matematica Italiana Publ.

[42] Emmer, M. (2020). "Homage to Octavia Spencer". In M. Emmer and M. Abate (eds.), *Imagine Math 7* (pp. 105–116). Switzerland: Springer Nature.

[43] *Dear Brigitte*, (1965). Filmmaker Henry Koster, with James Stewart, Bill Mumy, Glynis Johns, Brigitte Bardot, script by John Haase, Nunnally Johnson & Hal Kantor, USA.

[44] Emmer, M. and Manaresi, M. (eds.) (2003). *Mathematics, Art, Technology and Cinema*. Berlin: Springer verlag.

[45] *Fermat's Last Theorem*, (1996). Filmmaker Simon Singh, with Andrew Wiles and other mathematicians, script by Simon Singh and John Lynch, BBC production, UK.

[46] Singh, S. (2002). "L'emergere della narrativa scientifica". In M. Emmer (ed.), *Matematica e cultura 2002* (pp. 189–194). Milano: Springer.; English (ed.) (2005). *The Rise of Narrative Non-Fiction*. In M. Emmer (ed.), *Mathematics, Art and Architecture* (pp. 183–186). Berlin: Springer.

[47] *Venice 2008* (2008). Filmmaker Michele Emmer, M. Emmer and M. Villarreal prod., Rome, filmed during the 2008 conference.

[48] Emmer, M. and Abate, M. (eds.) (2020). *Imagine Math 7*. Switzerland: Springer Nature.

[49] Emmer, M. and Abate, M. (eds.) (2022). *Imagine Math 8*. Switzerland: Springer Nature.

[50] Emmer, M. (2020). "Soap bubbles Vanitas Venice". In: A. Wonders (ed.) *Math in the Time of Corona* (pp. 151–160). Mathematics Online First Collections. Switzerland: Springer Nature.

[51] Exhibition in Perugia: https://youtu.be/fFHh9hi5fwM; Soap Bubbles: https://youtu.be/iys6zVOMiqc.

https://doi.org/10.1142/9789811253072_0015

Chapter 14

Communicating Health Statistics Effectively: The Power of Icon Arrays and Natural Frequency Trees

Ines Lein*,†,‖ and Mirjam A. Jenny*,†,‡,§,¶

*Science Communication Unit, Robert Koch Institute, Berlin, Germany
†Harding Center for Risk Literacy, Faculty for Health Sciences Brandenburg, University of Potsdam, Germany
‡Max Planck Institute for Human Development, Berlin, Germany
§Science2Society, Health Communication, University of Erfurt, Erfurt, Germany
‖leini@rki.de
¶mirjam.jenny@uni-erfurt.de

Graphical representations can facilitate people's understanding of mathematical content. This is crucial, for example, when health statistics are communicated to patients. In this chapter, we describe different graphical means and will, in particular, focus on icon arrays and natural frequency trees. Icon arrays are already understood by children and made their way into school curricula. They can be used to explain the pros and cons of different medical interventions as well as their risks and are increasingly used in the media — albeit not often enough. Natural frequency trees allow for transparent representations of the effectiveness of medical tests and interventions.

Risks are often stated as percentages. This can lead to misconceptions, particularly when the presented number does not quantify a certain risk, but a *change* in risk. Statements such as "men over 6ft are twice as likely to get infected with COVID-19" [1], "belly fat in older women is linked to a 39% higher risk of dementia" [2], "breast cancer screening from age 40 leads to a 25% reduction in breast cancer mortality" [3] or "prisoners 550% more likely to get COVID-19" [4] are prevalent in the media, advertisements and even politics. Amidst the COVID-19 crisis in August 2020, the FDA announced at a press conference and subsequently via

Figure 1: FDA commissioner Steve Hahn on the microphone at a press conference in the White House on August 24, 2020.
Source: screenshot, https://twitter.com/US_FDA/status/1297662384060981248.

Twitter that a "35% improvement in survival is a pretty substantial clinical benefit", referring to the treatment of COVID-19 patients with convalescent plasma (Figure 1). The impressively large numbers cited in the examples above are relative changes and quantify relative risks. They should rather be presented transparently as absolute changes and absolute risks, but absolute numbers are usually far less impressive.

Beware of Relative Risk Changes!

We will illustrate the difference between absolute and relative risks with two examples in detail.

Our first example examines the warning from the World Health Organization that processed meat is carcinogenic [5]. It was found that eating 50 grams of processed meat a day would increase the chance of developing colorectal cancer by 18%. Looking at this relative risk increase of 18%, eating processed meat seems risky. However, this number leaves out two important risk aspects: the baseline risk that one develops colorectal cancer and the absolute risk increase caused by eating processed meat. A relative increase of 18% could (roughly) mean an increase from 500 in

1,000 people to 600 in 1,000 people that get diagnosed with the cancer, for example. However, it could also mean an increase from five in 1,000 people to six in 1,000 people. While the former risk increase would result in 100 additional diagnoses in 1,000 people, the latter would result in one additional diagnosis in 1,000 people. We cannot deduct the correct number of additional cancer diagnoses from the relative risk of 18% because crucial information is missing. Luckily, the latter example is correct: in absolute terms, the risk increase is one in 1,000 people, from five in 1,000 to six in 1,000 people. This information is more transparent than the 18% relative risk increase stated by WHO because it provides the base rate of how often the cancer occurs (five in 1,000 people get diagnosed with colorectal cancer) as well as the absolute risk increase that can be attributed to the consumption of processed meat (one additional diagnosis in 1,000 people). Stated in these absolute terms, eating processed meat no longer seems that risky.

Our second example highlights how much harm can be done if absolute and relative risk changes are not understood. In 1995, the British Committee for Drug Safety announced a warning that the new, third-generation birth control pill would double the risk of thrombosis — in other words, it would increase the risk by 100%. This number sounds extremely threatening at first glance and the media was quick to pick up the story. Many women were so worried by the news that they decided to stop taking the pill out of fear of thrombosis. This resulted in numerous unwanted pregnancies, which led to an estimated 13,000 additional abortions in England and Wales in the following few years [6], even though abortion rates had been steadily declining before the warning. Would absolute figures have had the same effect? The studies on which the warning was based showed that for every 7,000 women who took the second-generation pill, one woman developed thrombosis and that the number rose to two in women who took the third-generation pill. This means that the absolute risk increase was only one in 7,000, while the relative risk increase was indeed 100%. In absolute terms, however, this is a risk increase merely from 0.014% to 0.028%, i.e., by 0.014 percentage points. Put this way, the statistic probably would have worried women in Great Britain much less. Probably even less so when comparing the risk increase with the basic risk of thromboembolism, which can result from any intentional or unintentional pregnancy, which is approximately four in 7,000 and is thus larger [6]. If the women had known these numbers, many of them would probably not have stopped taking the pill. Many unwanted pregnancies and subsequent abortions could have been prevented, as well as the associated possible psychological and physical harm and the burden on the health care system. In summary, absolute risk change is more transparent and intuitive than relative risk change, since it carries more information. Relative risks can be derived directly from absolute risks — but this is not the case vice versa.

Visualizing Risks with Icon Arrays

A particularly nasty form of manipulation is mismatched framing. Here, the benefit of an intervention is expressed in terms of relative risks (in large numbers) and the damage in terms of absolute risks (in small numbers) [7]. Changes in risks and benefits can be visualized by means of icon arrays [8]. Icon arrays consist of a number (10 or 1,000 or 10,000 ...) of individual elements (dots or icons) representing a unit (people) within a population that allow for a visual as well as numerical comparison of quantities. They have been used in risk communication to convey evidence to facilitate health decisions, for example for cancer screenings and medical interventions.

Imagine you have to choose between two ointments against insect bites. There is a package of the product *sting-ex* in your cupboard that is well established and that you have used for years. But you heard of a new product called *no-itch* that was recommended to you by your pharmacist and is supposed to work 10% better: from the package leaflets you learn that *sting-ex* helps 10 in 100 people, while the new ointment *no-itch* cures one person more, i.e., 11 out of 100 people. The benefits of both drugs are visualized in Figure 2 using icon arrays that include all underlying quantities. Let's assume that the new medication unfortunately has an unpleasant side effect: Some people experience a rash and have to stop using the ointment immediately. The side effect also occurs with the established *sting-ex* but somewhat less frequent. The pharmacist mentioned

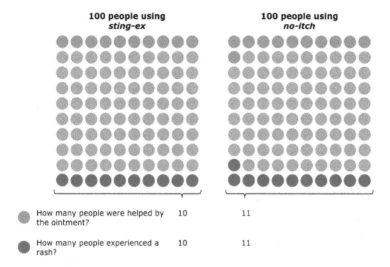

Figure 2: Visualizing the benefits and harms of the fictional drugs *sting-ex* and *no-itch* using icon arrays. The reference class is 100 people using the medication. The absolute increase in the chance of healing is 1 percentage point; the relative increase in the chance of healing is 10%. The absolute increase in the chance of side effects is 1 percentage point, the relative increase is 10%.

an increased risk for a rash of only 1%. Flipping through the package leaflet you read that 10 out of 100 people suffered the side effect in the old *sting-ex*, in *no-itch* it is 11 people; indeed the risk increase is one in 100 people (1%). This is also visualized in Figure 2. In this simplified example, the pharmacist used mismatched framing to sell the new product. Incidentally, such non-transparent arguments are actually used to facilitate the approval of new drugs or to communicate health risks [9].

Of course, icon arrays can also be created for other sample sizes and there are simple online tools available, for example at www.iconarray.com, as well as R packages that assist with their design [10].

Because many people have difficulties using numbers and processing elementary probability expressions, icon arrays are a promising way of communicating medical risks to a wide range of target groups. Galesic, Garcia-Retamero, and Gigerenzer examined whether people with high- and low-numeracy skills can understand risk numbers more accurately when they are displayed in icon arrays [11]. They found that icon arrays generally improved the accuracy of estimated relative risk reductions when the presented icon array was accompanied by numerical descriptions. Icon arrays were particularly useful for study participants whose numeracy skills were lower than the median of their respective group. This is backed up by other research that shows that well-designed visual aids like icon arrays are useful for people who may have problems understanding information about health risks, for example people with low-numeracy skills, low knowledge about medical facts or both, older adults, and immigrant populations [12–14]. A systematic review that looked at different formats of visual aids that promote risk literacy in the health domain came to the conclusion that "on the basis of all relevant available data detailed in this review, we find that icon arrays tend to be the best 'all purpose' type of visual aids" [15]. The authors suggest applying a "less is more" approach for the design as very simple icon arrays convey the meaning of important information best for all target groups. Ideally, five simple rules should be followed to communicate a risk reduction or the risk of side effects [15]: 1) use different icon arrays to represent the results of the baseline risk (i.e., for the sample of non-treated individuals) and the incremental/reduced risk due to treatment (i.e., for the sample of treated individuals); 2) depict both, the affected individuals and the entire population at risk; 3) keep the size of the entire population at risk in the treated and non-treated groups of individuals constant for comparison; 4) arrange icon arrays as groups in a block rather than in a random scattering; and 5) use person-like icons or more abstract icons (dots, squares).

Very good examples of icon arrays for various medical interventions and screenings are disseminated on the website of the Harding Center for Risk Literacy (https://www.hardingcenter.de/en) where the need for transparent, patient-friendly risk communication in health care is highlighted. The site also features a large array of 10,000 icons that compares the estimated numbers for COVID-19 infections and deaths in Germany — converted to a football stadium — with past disease outbreaks and various everyday hazards (Figure 3). Icon arrays are also finding their way to commercial providers of health services to illustrate proportions. An example [16] is the reports of the American personal genomics and biotechnology company 23andMe shown in Figure 4. The icon array displays the chances of a certain personal trait outcome

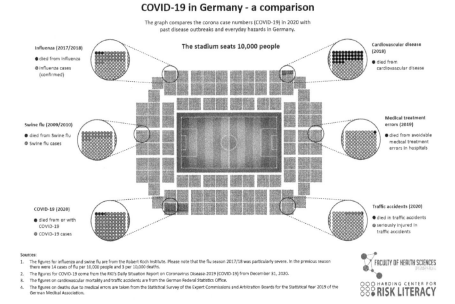

Figure 3: Image comparing cases and fatalities from different disease outbreaks and everyday hazards.
Source: Graphic provided by Harding Center for Risk Literacy.

Figure 4: Excerpt from the results of a genetic test provided by the direct-to-customer genetic testing company 23andMe, showing icon arrays illustrating the chances of a certain personal trait outcome.
Source: https://blog.23andme.com. © 23andMe, Inc. 2022. All rights reserved and distributed pursuant to a limited license from 23andMe.

relating to the customers' ancestry and genetic predisposition. Simple icon arrays are even suitable to teach children about percentage points in 10 x 10 grids [17].

Conveying Risk Numbers in Tables: Fact Boxes

Like icon arrays, fact boxes communicate risks transparently and present the best available evidence on a medical topic as they are developed according to the methods of evidence-based medicine [18,19]. They show a balanced overview of benefits and harms by comparing at least two groups of equal size in a table format. Fact boxes consist of six main features: 1) a summary statement that describes the benefits and harms of a medical treatment without making a recommendation; 2) exhaustive and explicit information about the reference population (e.g., gender, age group) and any additional information that may influence the understanding of the reported facts to avoid ambiguity of the interpretation of the content; 3) explicitly stated benefits and harms of the intervention in the form of a list of statements or questions; 4) a comparison of results between a treatment and a control group based on scientific evidence (from meta-analyses, systematic reviews, or randomized controlled trials); 5) a measure of the effect for each group, presented as absolute numbers out of a total sample of, for example, 100 or 1,000 people; and 6) the sources for all information and the date of the last update [20,21].

Due to their clean design, fact boxes help even people with low medical and statistical literacy to make people to make informed decisions. Several studies show that fact boxes are effective tools for informing the general public about the harms and benefits of medical interventions [22–25]. Figure 5 shows a fact box for a COVID-19 mRNA vaccine for adults under the age of 60.

Natural Frequencies and Their Graphical Representation as Trees

Bayesian inference tasks are easier to solve when they are represented as joint frequencies obtained through natural sampling, known as natural frequencies, than as conditional probabilities. Therefore, natural frequencies can help experts and laypeople better understand medical

Fact box: How safe and effective are COVID-19 mRNA vaccines for adults under the age of 60?

This fact box compares adults under the age of 60 years who have not been vaccinated against COVID-19 (left side) with vaccinated adults (right side). It is assumed that 240 out of 1,000 unvaccinated people will get sick. This is comparable to your risk of getting sick if you have close contact with someone who is infected.

	1,000 non-vaccinated adults	1,000 vaccinated adults
Benefits of the vaccine		
How many get COVID-19?	240	10
... and – depending on their age or previous medical conditions – have to be treated in hospital due to severe illness?	6 to 31	0 to 1
... and experience long-term complications from COVID-19?	The numbers are still uncertain. There is clear evidence of permanent respiratory distress and memory impairment due to COVID-19 ("long COVID").	
Harms from the vaccine		
How many are unable to participate in their daily activities (due to temporary fatigue, fever, aches, or chills) on individual subsequent days due to a vaccine dose?	0	82
How many suffer severe harm (e.g., allergic overreaction) within a month due to a vaccine dose?	0	Close to 0
How many suffer from long-term complications due to the vaccination?	0	There are currently no indications of long-term complications.

Note: Typical vaccine reactions, which may affect the arm or the entire body, usually subside after one to two days. The occurrence of rare vaccine reactions, such as allergic reactions, and potential relationships between the vaccine and less typical reactions (e.g., insomnia, enlarged lymph nodes, and transient facial paralysis) are currently under examination. It is not yet clear how long the vaccine provides protection.

Sources for the vaccines Comirnaty (manufacturer BioNTech/Pfizer) and Moderna (manufacturer Moderna): Baden 2020. NEJM; BioNTech & Pfizer 2020. www.comirnatyeducation.de; CDC 2021. MMWR; EMA 2020. www.ema.europa.eu; FDA 2020. FDA Briefing Document; Polack 2020. NEJM; RKI 2020. reporting data; STIKO 2021. Epidemiological Bulletin.

Figure 5: Fact box to illustrate the benefits and harms of a COVID-19 mRNA vaccine for adults under the age of 60. All endpoints were calculated to the denominator of 1,000 adults this age. Out of 1,000 non-vaccinated adults, approximately 240 get COVID-19, compared to 10 in 1,000 vaccinated adults. From 6 to 31 out of 1,000 non-vaccinated adults under the age of 60 have to be treated in a hospital due to severe illness, whereas it is only 0 to 1 out of 1,000 vaccinated adults under the age of 60. Out of 1,000, 82 adults under 60 are unable to participate in their daily activities due to temporary fatigue, fever, aches, or chills on individual subsequent days due to a vaccine dose. Note that these numbers may change as the virus evolves.
Source: https://www.hardingcenter.de/en/fact-boxes-on-the-mrna-vaccine-against-covid-19.

test results or screenings and are useful for communicating the central concepts of medical diagnostics [26]. As detailed in Figure 6, natural frequencies can be visually represented as natural frequency trees. These consist of absolute frequencies obtained through natural sampling. Often, specifics for medical tests are given in conditional probabilities that are difficult to process for most people, including physicians. Two important probabilities can be calculated with the help of natural frequency trees: First, the probability that a person who receives a positive test result actually has the disease (positive predictive value, PPV). Second, the probability that a person who receives a negative test result does not have the disease (negative predictive value, NPV). As an example, we would

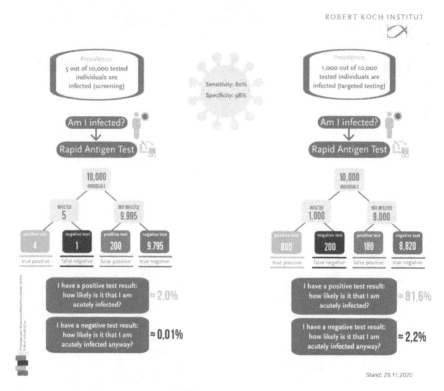

Figure 6: Natural frequency trees comparing SARS-CoV-2 rapid antigen test results for the two different test strategies screening (left) and targeted testing of individuals with COVID-19 symptoms (right). The figure shows the assumed test specifications sensitivity (80%) and specificity (98%), the proportion of tested people who are infected (five out of 10,000 and 1,000 out of 10,000 tested people are actually acutely infected), and the positive and negative predictive values for both test strategies.

like to illustrate how a natural frequency tree helps to understand rapid antigen tests results for SARS-CoV-2.

Antigen detection diagnostic tests are designed to directly detect SARS-CoV-2 proteins produced in respiratory secretions by the replicating virus. By September 2020, nearly 100 companies were developing or manufacturing rapid tests for SARS-CoV-2 antigen detection [27]. Even in situations in which case numbers are too high to contain the virus, most people in the population are not, or no longer, infected. Natural frequency trees help illustrate how heavily the expected numbers of correct and incorrect test results depend not simply on the test quality criteria sensitivity and specificity but on the proportion of tested people who are in fact infected (i.e., the prevalence or pre-test probability). The sensitivity of a test is the proportion of tested infected people with a positive test result, i.e., the number of infected people with correct positive results divided by the number of tested people with the infection. The specificity of a test is the proportion of tested non-infected people with a negative test result, i.e., the number of non-infected tested people with correct negative results divided by the number of non-infected tested people. Figure 6 compares test results for two different hypothetical situations: one in which a small proportion of those tested is infected (e.g., in untargeted testing of the total population) and another in which a large proportion of those tested is infected (e.g., in targeted testing of individuals with COVID-19 symptoms). In the example of untargeted testing, this proportion is approximately five out of 10,000 people (left), while in the example of targeted testing based on their COVID-19 symptoms, it is about 1,000 out of 10,000 people (right).

As mentioned above, the positive predictive value quantifies the probability that a person who receives a positive test result is actually infected with the virus. This means for our examples in Figure 6: $4/(4 + 200) \approx 0.02$ for untargeted testing and $800/(800 + 180) \approx 0.816$ for targeted testing. Depending on the test strategy, the PPV is 2.0% (untargeted testing) or 81.6% (targeted testing).

The negative predictive value quantifies the probability that a person who receives a negative test result is not infected with the virus. For the example in Figure 6: $9,795/(9,795 + 1) \approx 0.9999$ (99.99%) for targeted testing and $8,820/(8,820 + 200) \approx 0.978$ (97.8%) for untargeted testing. Depending on the test strategy, the probability that a person is infected despite a negative test result is hence 0.01% (untargeted testing) or 2.2% (targeted testing). Note that the positive predictive value and the negative predictive value do not add up to one and 100%, respectively, as they refer to different groups (people with positive test results and people with negative test results).

These natural frequency trees are targeted at a professional audience to help them with the interpretation of test results during the SARS-CoV-2 pandemic. Together with an interactive tool, they were published on the website of the Robert Koch Institute, the national public health institute responsible for disease control and prevention in Germany [28].

Conclusion

We summarize that relative risk numbers can lead to an overestimation of medical risks or benefits of medical interventions. Therefore, benefits and harms should always be presented as absolute risks or the absolute risks should at least accompany the relative risk numbers. Icon arrays and fact boxes help to communicate the advantages and disadvantages of medical interventions and should be preferred over plain text for communicating numerical harms and benefits. The graphical representation of natural frequencies as trees support the understanding of medical test results, for example, the positive and negative predictive values.

Even though icon arrays, fact boxes, and natural frequency trees look simple, developing them often requires a considerable amount of time and work that includes systematic literature reviews and transformation of the results into a transparent message. Informed decisions are only possible when individuals understand the potential harms and benefits of different choices. It is therefore fortunate that the advantage of these simple formats are being increasingly recognized in public health. Especially during the COVID-19 pandemic, people need to be supplied with reliable and easy to understand information. We hence strive to continually develop new materials to inform the public about the benefits and harms of different health interventions, about continuous and emerging health threats, and about preventive and protective measures. To this end, we often collaborate with other experts or organizations, such as the Harding Center for Risk Literacy, the University of Erfurt, or the German Federal Center for Health Education (BZgA).

References

[1] Chalmers, V. (July 28, 2020). "Does being tall raise the risk of getting Covid-19? Men over 6 ft are twice as likely to get infected, study claims". *Daily Mail.* https://www.dailymail.co.uk/news/article-8568125/Men-6ft-TWICE-likely-infected-Covid-19-study-claims.html. [Accessed on September 19, 2020].

[2] Prior, R. (June 23, 2020). "Belly fat in older women is linked to a 39% higher risk of dementia within 15 years, study says". *CNN.* https://edition.cnn.com/2020/06/23/health/belly-fat-dementia-link-wellness/. [Accessed on September 19, 2020].

[3] Hospital Healthcare Europe. (August 17, 2020). "Earlier mammograms associated with a reduction in breast cancer mortality". https://hospital healthcare.com/clinical/pathology-and-diagnostics/earlier-mammograms-associated-with-a-reduction-in-breast-cancer-mortality/. [Accessed on September 19, 2020].

[4] Sternlicht, A. (July 8, 2020). "Prisoners 550% more likely to get Covid-19, 300% more likely to die, new study shows". *Forbes*. https://www.forbes.com/sites/alexandrasternlicht/2020/07/08/prisoners-550-more-likely-to-get-covid-19-300-more-likely-to-die-new-study-shows/. [Accessed on September 30, 2020].

[5] International Agency for Research on Cancer. (2015). "Q&A on the carcinogenicity of the consumption of red meat and processed meat". *World Health Organization*. https://www.iarc.who.int/wp-content/uploads/2018/11/Monographs-QA_Vol114.pdf. [Accessed on September 23, 2020].

[6] Furedi, A. (1999). "Social consequences: The public health implications of the 1995 'pill scare'". *Hum Reprod Update*, 5(6), 621–626.

[7] Sedrakyan, A., and Shih, C. (2007). "Improving depiction of benefits and harms: Analyses of studies of well-known therapeutics and review of high-impact medical journals". *Medical Care*, 45, 523–528.

[8] Binder, K., Krauss, S., and Gigerenzer, G. (2020). "Risikoveränderungen: Wie absolute und relative Veränderungen von Risiken mit Bildgittern unterrichtet werden können". *Mathematik lehren*, 220, 12–15.

[9] Gigerenzer, G. (2015). *Risk Savvy: How to Make Good Decisions* (1st edn.). New York: Penguin.

[10] Gilbey, L. "ggwaffle R package". https://liamgilbey.github.io/ggwaffle/.

[11] Galesic, M., Garcia-Retamero, R., and Gigerenzer, G. (2009). "Using icon arrays to communicate medical risks: Overcoming low numeracy". *Health Psychology*, 28(2), 210.

[12] Garcia-Retamero, R., and Cokely, E. T. (2013). "Communicating health risks with visual aids". *Current Directions in Psychological Science*, 22, 392–399.

[13] Garcia-Retamero, R., and Cokely, E. T. (2014). "Using visual aids to help people with low numeracy make better decisions". In B. L. Anderson and J. Schulkin (eds.), *Numerical Reasoning in Judgments and Decision Making about Health* (pp. 153–174). Cambridge, UK: Cambridge University Press.

[14] Garcia-Retamero, R., and Dhami, M. K. (2011). "Pictures speak louder than numbers: On communicating medical risks to immigrants with limited non-native language proficiency". *Health Expect*, 14(s1), 46–57.

[15] Garcia-Retamero, R., and Cokely, E. T. (2017). "Designing visual aids that promote risk literacy: A systematic review of health research and evidence-based design heuristics". *Human Factors*, 59(4), 582–627.

[16] Krock, B. (February 27, 2018). "Your traits never looked better". *23andMeBlog*. https://blog.23andme.com/health-traits/traits-never-looked-better/. [Accessed on September 30, 2020]. © 23andMe, Inc. 2022. All rights reserved and distributed pursuant to a limited license from 23andMe.

[17] Staatsinstitut für Schulqualität und Bildungsforschung (ISB) (2018). "LehrplanPlus Gymnasium Bayern Mathematik 6. Klasse". https://www.lehrplanplus.bayern.de/fachlehrplan/gymnasium/6/mathematik. [Accessed on September 19, 2020].

[18] Kunz, R., Ollenschläger, G., Raspe, H., Jonitz, G., and Donner-Banzhoff, N. (2007). *Lehrbuch evidenzbasierte Medizin* (2nd edn.). Köln: Deutscher Ärzteverlag.

[19] Lühnen, J., Albrecht, M., Mühlhauser, I., and Steckelberg, A. (2017). *Leitlinie evidenzbasierte Gesundheitsinformation*. Hamburg. https://www.ebm-netzwerk.de/de/medien/pdf/leitlinie-evidenzbasierte-gesundheitsinformation-fin.pdf. [Accessed on November 6, 2020].

[20] Hinneburg, J., Wilhelm, C., and Ellermann, C. (2020). Methods Paper for Developing Fact Boxes, Version 2.2. Harding Center for Risk Literacy (Hrsg.), University of Potsdam, Faculty of Health Sciences. https://www.hardingcenter.de/sites/default/files/2020-methods_ paper_Harding-Center_EN.pdf. [Accessed on November 6, 2020].

[21] McDowell, M., Gigerenzer, G., Wegwarth, O., and Rebitschek, F. G. (2019). "Effect of tabular and icon fact box formats on comprehension of benefits and harms of prostate cancer screening: A randomized trial". *Medical Decision Making*, 39(1), 41–56.

[22] Brick, C., McDowell, M., and Freeman, A. L. (2020). "Risk communication in tables versus text: A registered report randomized trial on 'fact boxes'". *Royal Society Open Science*, 7(3), 190876.

[23] Loizeau, A. J. Theill, N., Cohen, S. M., Eicher, S., Mitchell, S. L., Meier, S. ... Riese, F. (2019). "Fact box decision support tools reduce decisional conflict about antibiotics for pneumonia and artificial hydration in advanced dementia: A randomized controlled trail". *Age Ageing*, 48(1), 67–74.

[24] McDowell, M., Rebitschek, F. G., Gigerenzer, G., and Wegwarth, O. (2016). "A simple tool for communicating the benefits and harms of health interventions: A guide for creating a fact box". *MDM Policy Pract*, 1(1), 2381468316665365.

[25] Schwartz, L. M., Woloshin, S., and Welch, H. G. (2009). "Using a drug facts box to communicate drug benefits and harms two randomized trials: Two randomized trials". *Annals of Internal Medicine*, 150(8), 516–527.

[26] McDowell, M., and Jacobs, P. (2017). "Meta-analysis of the effect of natural frequencies on Bayesian reasoning". *Psychological Bulletin*, 143(12), 1273.

[27] World Health Organization. (2020). "Antigen-detection in the diagnosis of SARS-CoV-2 infection using rapid immunoassays: Interim guidance". https://www.who.int/publications/i/item/antigen-detection-in-the-diagnosis-of-sars-cov-2infection-using-rapid-immunoassays. [Accessed on November 8, 2020].

[28] Robert Koch-Institut, "Corona-Schnelltest-Ergebnisse verstehen". https://www.rki.de/DE/Content/InfAZ/N/Neuartiges_Coronavirus/Infografik_Antigentest_Tab.html. [Accessed on February 5, 2021].

© 2023 World Scientific Publishing Co. Pte. Ltd.
https://doi.org/10.1142/9789811253072_0016

Chapter 15

How to Develop a Concept of Smart-World Literacy: A Roadmap

Felix G. Rebitschek

Harding Center for Risk Literacy,
Faculty of Health Sciences Brandenburg, University of Potsdam, Germany;
Max Planck Institute for Human Development, Berlin, Germany
rebitschek@uni-potsdam.de

Societal and economic processes between organizations and individuals are increasingly designed with algorithms. Examples are digital consumer scorings, digital nudges, and dark patterns in marketing. These algorithm-based information architectures, more precisely their design, their interaction with individuals and the resulting dynamics, can empower but also manipulate. Therefore, political as well as civil society and consumer-specific debates are necessary. However, to take part in these debates and in novel societal and economic processes, individual prerequisites are required. Various approaches to algorithm literacy and data literacy (besides media literacy, digital literacy, competencies in the digital world, information literacy, etc.) attempt to define these prerequisites, but ignore the overlap with existing research too often. Taking into account the concepts that seek to formalize algorithm-specific preconditions, a roadmap is presented in order to inform developers of literacy concepts for the smart world.

The roadmap: From an existing general competence framework (e.g., concept of data literacy), competence fields (e.g., planning) can be isolated for the area of algorithms by analyzing the fields' characteristics (including different competencies, for example, information search). On the one hand, it must be examined which existing concepts address these fields of competence at least in part. Overlaps between different concepts have to be identified and demarcations need to be made to the general competence framework. On the other hand, it has to be determined whether the current real-world state of the challenge is addressed. To this end, a test scheme has to be formulated to assess whether one's competence framework of existing (and possibly added) competence fields can be ecologically valid. If the selection is justified, the premises — the motivational selection — need to be made explicit. One example is the premise that consumers can obtain quality-assured results when facing algorithms. Such premises are needed to determine literacy objectives (example: consumer has the competence to act with regard to receiving quality information by his scoring provider). Only with these objectives, the literacy levels can be formalized. The levels can range from awareness of systems to a commitment to algorithm regulation. It is then important to specify the competence contents implied by the chosen competence fields and the subject matter of algorithms (e.g., fairness of a classification) and to sharpen them to such an extent that quantitative investigations are possible. The level assignment has to be reproduced reliably with the selected contents in empirical tests. Standardization studies would then be necessary. In addition to such efficacy proofs in assessment, randomized controlled (experimental) studies should prove the effectiveness of competence differences. Qualitative work is also needed here to explore the scope of competence differences, for example with regard to trust. The roadmap presented is purpose-oriented to substantiate smart-world literacy concepts with regard to education and regulation needs but also innovation potentials.

This text from the psychological perspective is supposed to support researchers and practitioners who aim at assessing people's capabilities to navigate the digital world. This roadmap illustrates essential steps that help conceptualize a set of applicable characteristics and skills, simplistically referred to here as "smart-world literacy". Yet, smart-world literacy itself is not presented as a literacy concept here. It remains in the hand of the researchers and practitioners, which competences they aim to develop.

Introduction

One could develop the propensity to underestimate the impact of algorithmic technologies, when facing the permanent accumulation of superficial promises and buzzwords (e.g., digital nudges, dark patterns, and people analytics). In particular, buzzwords can indicate the absence of qualitative evidence in terms of effectiveness, and economic or societal progress.

Digital nudges as the instances of a "choice architecture that alters people's behavior in a predictable way without forbidding any options or significantly changing their economic incentives" [1] can convey the ultimate preference of the designer of the architecture [2], but do not support informed decision-making of citizens regularly (for a discussion, see [3]). Dark pattern designs [4] like hidden opt-out options, that make you buy things you do not want, recycle what companies' advertizing departments and agencies did for a 100 years.

People analytics that screen applicants or score employees lack rigorous proofs of being beneficial [5], for instance when they miss opportunities (false negatives). The gist is, there is huge uncertainty as well as speculation bubbles on the efficacy and effectiveness of information architectures that exploit data about individuals. At once, one can hardly overestimate their impact in terms of societal change, because expectations are selling, with positive and negative consequences. One can observe how information architectures reshape processes between organizations and individuals; this is Airbnb in accommodation [6], also this is a scoring of pupils for future public resource allocation [7].

Algorithm-based information architectures of organizations, the way in which they interact with the individuals, and the resulting dynamics could theoretically empower the individuals (new learning opportunities and qualities) but also manipulate (motivated information provision) (e.g., [8]).

So, political discourse as well as civil society and consumer-specific debates about those architectures, and of their structural and process design regarding interactions and dynamics are essential. But who can take part in these discourses and shape them? Individual features, skills, and attitudes are required, but which [9]? The potential answer is subject to manifold theoretical approaches: to digital literacy as the "ability to understand and use information in multiple formats from a wide range of sources" (Gilster, 1997, p. 1 in [10]), to critical digital literacy about analyzing and judging the content, usage and artefacts of technology while considering how meaning is constructed, by whom and for what purposes [11], to digital competence involving "the confident and critical use of Information Society Technology (IST) for work, leisure and communication" (European Commission, 2006, p. 16 in [12]), to media competences that assemble skills and abilities with regard to the use of media under the general premise of constructive participation in society [13], and to information literacy ("Can I trust [given] information" and "Which information do I publish?", [14]). Most aptly, algorithm literacy and the ability to deal with data in a planned manner and to consciously use and question it in the respective context (data literacy, [15]) seem to assemble those prerequisites. However, the research field — which in its core wants to investigate differential capabilities to deal with the algorithmic data-driven world of novel digital media — is very heterogeneous and marked by isolated concepts that neither connect much to each other nor to the real world [16].

This chapter aims to provide researchers and practitioners with a roadmap to compile, select, and evaluate the individual prerequisites in a more standardized way. The roadmap shows how one might develop a concept of literacy, which is useful for the digital so-called smart world. Starting with a clarification of the goals of the conceptual development, the selection of a competence framework with competence fields as well as the motivation behind it need to be justified, before defining literacy objectives and levels within the competence fields and empirically demonstrating their relevance.

At the Starting Point: Clarify the Goals

The presented roadmap is pragmatic in that clarification of purpose is seen as the starting point of any smart-world literacy effort. Why is smart-world literacy necessary? What can we achieve with such a concept? Pedagogical interventions that target individual prerequisites for dealing with information architectures represent a typical goal. A competence framework — as a bridge to a formalized description — that could help concretize education provides benchmarks for evaluating the intervention of interest. For instance, do employees need to complete specific training modules on algorithms for a particular career path? Does an average grade from a course on data literacy give access to a particular university? Another goal is to identify regulatory gaps. A competency framework helps identify requirements for regulatory action by providing information about the limited external effectiveness even of perfect competency. For example, what kind of vulnerabilities in a digital device can be addressed by a trained computer scientist, and what cannot be addressed? Another exemplary goal is to foster innovation potential. A competency framework for selecting and developing the creativity and progress of individuals or teams makes explicit demands on the methodology used to achieve improvements.

Finally, the predominant goal of an assessment is to describe a population, a group, or an individual status in a meaningful way. A meaningful competency framework enables integrative work that helps differentiate existing theories and models, and implies external validation with real-life variables. For the following outline of the roadmap along the line with different branches, the goal is an assessment with high external validity for groups or populations. The case for individual diagnostics, a diagnostic inventory of smart-world literacy, needs to be further elaborated in the future. Difficulties here arise from our lack of knowledge about individual factors and cognitive preconditions related to personal smart-world "success", which would be necessary to design differentiating standardization studies.

First Branch: Choose a Competence Framework

With regard to model integration, existing general competence frameworks help in a first step to isolate relevant competence fields for smart-world literacy. Competence frameworks generally and comprehensibly describe effective acting in a defined context (competence fields), including competence definitions and embedding societal values, objectives of stakeholders, and also demarcations. Competence frameworks are helpful because they allow for competence formalizations that organizations could make use of. The literature search should at least have the character of a reproducible quick systematic search [17] with predefined data sources, search terms (on the goal of the concept, competence frameworks), synonyms, and publication types (e.g., theoretical or systematic review) within a technology-adequate time period. Ideally, an up-to-date competence framework with competence fields, competences with their objectives, and competence levels is found to have been built on a systematic review (as an example, see the competence framework for data literacy by Schüller, Busch, and Hindinger, 2019 [16]).

If the competence fields included in a particular competence framework (= the addressed context) are sufficient for the requirements of one's conceptual goals, one can continue to use that particular framework and the implied competence fields. However, it is more likely that relevant competence fields are included and others are missing, or that there are several competence frameworks with the same objective. In the latter case, the quality of the evidence is a helpful yardstick for selecting a competence framework (e.g., based on a systematic review that includes assessments of the empirical evidence). Nevertheless, when faced with a choice between different competency frameworks, it is preferable to list the respective competency fields and compare them with one's own notions of relevant fields (e.g., application contexts). Pragmatically, one chooses the framework with the most similar competence fields (provided that the theory behind it is convincing).

Second Branch: Modify the Competence Framework

This step should address (from the researcher's or practitioner's perspective) the problem of considering a specific competence field that is neglected by the chosen competence framework. Please note that the way in which researchers and practitioners collect their own set of candidate competence fields — e.g., from experience (e.g., exemplary observations of consequences from human-algorithm interaction; collecting real smart-world use cases) or with the help of expert workshops — is not further explained in this chapter. In order to tie in with existing research, it is necessary to investigate which concepts in the literature at least partially address the context of interest (competence field). Again, a rapid systematic search with search terms and synonyms of possible competences in the relevant competence field is recommended. For example, if understanding statistical data (competence field) is relevant to the desired competence framework but is not included in existing frameworks, such a search would reveal the concept of risk literacy. Risk literacy is a set of competencies that enable a person to search for, think about, evaluate, and act on statistical evidence for their own benefit or that of others (e.g., [18]). If this concept is included, it would enrich the chosen competence framework. By adding concepts, the overlaps and differences between the competence fields within the framework have to be worked out.

In a final step, it needs to be explored if the state of the real-world challenge (goals) can be addressed by this competence framework. For example, if the skills to interact with systems of algorithm-based behavior prediction and control (scoring) are to be assessed [19], technological features (algorithm input features, output and outcomes, performance, fairness, etc.) may need to be examined to see if it is necessary or beneficial to understand them. To do this, a testing scheme needs to be formulated to check whether one's framework of competence fields might be ecologically valid. The simple question is: How do I recognize that differences in smart-world literacy matter? For example, a perfect score in smart-world literacy can be irrelevant in terms of real-world outcomes. Effective differences could be indicated with gradual validity: from unintentional to intentional sharing of personal information under uncertainty (behavior), receipt of personalized information, recommendations and offers (e.g., advertizing, prices, tariffs, social contacts), to hard outcomes such as experiences of social isolation, reduced participation, negative financial, or health outcomes. A checklist of effective differences needs to be compiled.

Third Branch: Justify and Select Literacy Objectives

When a competence framework — the formal and factual selection of competence fields — is defined, the premises — the motivational definition of competences — must be made explicit. Motivations can be political, ethical or result from social processes (often mediated by mass media). An example is the aspiration that consumers who are competent in digital information seeking can achieve quality-assured results. The premise here is informed decision-making, according to which consumers are empowered to weigh possible benefits and harms when pursuing decision options. Theoretically, one could alternatively propose that a competent consumer finds information quickly, or that he/she finds information from government sources safely. The researcher or practitioner must therefore reflect on and justify the premises underlying each competence. Since this process is subject to personal evaluation, taking into account the general competence goals, the premises, and their justifications should be presented transparently.

Subsequently, literacy objectives (in German "Kompetenzziele") need to be derived from the premises. In our example with the premise of informed decision-making, a literacy objective may be a consumer's successful action to obtain evidence of information quality from an information provider. He/she is aware of varying information quality online, recognizes the lack of evidence, and knows how to ask whom. The example implies that the definition of smart-world literacy objectives can be based on the analysis of building blocks which are required to address typical challenges.

Fourth Branch: Formalize Levels of Literacy

Based on smart-world literacy objectives, levels of literacy can be formalized. These levels represent the different standards that could or should be achieved. For our example of obtaining evidence of informational quality from an information provider, one can imagine a range of standards from problem awareness (e.g., low literacy level in information search), to evaluation of information offers (e.g., medium literacy level in contextualizing found information), to reflections on policy or regulatory activities about digital information quality (e.g., higher literacy level in maximizing benefits for others and oneself). Any classification of groups (or individuals) to different levels of literacy must be reliably reproducible in empirical tests with similar groups, ideally with standardization studies on differing individuals.

Prior to that, however, the content implied by the selected fields of competence must be selected for the object of the smart world (e.g., fairness of a classification) and formally sharpened to such an extent that quantitative tests are possible. This implies that items must also be developed for each concept of the individual prerequisites in order to collect the responses of the individuals that prove the relevance of the concept.

Fifth Branch: Prove Effectiveness of Differences

For item development, testing is the key mechanism. Self-report items should be avoided if possible (for a critical equivalent, see discussions about Health Literacy; [20]). Items intended to capture a competency such as smart-world literacy should, after extensive pretesting for comprehensibility, go through a process of scale development that reduces the item pool according to item discriminability and difficulty (e.g., based on the Rasch model; [21]). Depending on the level of the specified competencies, confirmatory and additional exploratory factor analyses help obtain an inventory that captures smart-world literacy. Assessments with the selected item set should be used together with alternative similar inventories to investigate discriminative aspects.

Special attention needs to be paid to external validity, which can be seen as the vanishing point of the whole proposal in this chapter. In order to demonstrate the relevance of a literacy approach in terms of effectiveness, one needs to refer to the checklist for effective differences from above. Hard outcomes such as experiences of social isolation, reduced participation, negative financial or health outcomes are on the one hand rather difficult to measure reliably (e.g., self-report), on the other hand the external validity of actual differences in smart literacy might be too weak. This means that inter-individual differences produce only very small effects in positive or negative outcomes, which make their evaluation difficult or impossible. However, data exchange behavior, that is interaction behavior with information environments, can be tested or at least queried with competence and knowledge items.

Besides such evidence of efficacy for population and group assessments, randomized controlled (intervention) studies could later prove the effectiveness of competence differences. Not to be neglected here is also qualitative work to characterize the quality of competence differences, for example, in relation to trust. In focus groups with members of differential competence, use cases should be discussed and perceived hurdles of items could be collected with the help of cognitive interviews.

Limitations and Conclusion

What is not presented in detail in this text are procedures for assessing candidate competences outside the scientific literature. Scale development for differential assessments [22] is also only mentioned in this theoretical work. Each step described in this chapter is applied in existing contexts, but whether the application of the proposed work process provides reliable tools or interventions to help navigate the smart world cannot be answered today. Obviously, empirical confirmation is needed at this point (e.g., regarding understanding of algorithm performance [23]), ideally in competition with existing scales that target the smart world. Only in this way can the present roadmap help to compile, select, and evaluate inter-individual differences in an evidence-based manner.

References

[1] Thaler, R. H., and Sunstein, C. R. (2009). *Nudge: Improving Decisions about Health, Wealth, and Happiness.* Penguin.

[2] Lembcke, T. B., Engelbrecht, N., Brendel, A. B., and Kolbe, L. (2019). "To nudge or not to nudge: Ethical considerations of digital nudging based on its behavioral economics roots". In *Proceedings of the European Conference on Information Systems (ECIS)*, Stockholm and Uppsala, Sweden.

[3] Blumenthal-Barby, J. S. (2013). "Choice architecture: A mechanism for improving decisions while preserving liberty?". In C. Coons and M. E. Weber (eds.), *Paternalism. Theory and Practice* (pp. 178–196). Cambridge: Cambridge University Press.

[4] Mathur, A., Acar, G., Friedman, M. J., Lucherini, E., Mayer, J., Chetty, M., and Narayanan, A. (2019). "Dark patterns at scale: Findings from a crawl of 11K shopping websites". *Proceedings of the ACM on Human-Computer Interaction*, 3(CSCW), 1–32.

[5] Marler, J. H., and Boudreau, J. W. (2017). "An evidence-based review of HR analytics". *The International Journal of Human Resource Management*, 28(1), 3–26.

[6] Garcia-López, M. À., Jofre-Monseny, J., Martínez-Mazza, R., and Segú, M. (2020). "Do short-term rental platforms affect housing markets? Evidence from Airbnb in Barcelona". *Journal of Urban Economics*, 119, 103278.

[7] Kelly, A. (2021). "A tale of two algorithms: The appeal and repeal of calculated grades systems in England and Ireland in 2020". *British Educational Research Journal*.

[8] Friedewald, M., Lamla, J., and Roßnagel, A. (eds.). (2017). *Informationelle Selbstbestimmung im digitalen Wandel.* Springer Fachmedien Wiesbaden.

[9] Eshet-Alkalai, Y. (2004). "Digital literacy: A conceptual framework for survival skills in the digital era". *Journal of Educational Multimedia and Hypermedia*, 13(1), 93–106.

[10] Pangrazio, L., Godhe, A. L., and Ledesma, A. G. L. (2020). "What is digital literacy? A comparative review of publications across three language contexts". *E-Learning and Digital Media*, 17(6), 442–459.

[11] Hinrichsen, J., and Coombs, A. (2013). "The five resources of critical digital literacy: A framework for curriculum integration". *Research in Learning Technology*, 21, 1–16.

[12] Spante, M., Hashemi, S. S., Lundin, M., and Algers, A. (2018). "Digital competence and digital literacy in higher education research: Systematic review of concept use". *Cogent Education*, 5(1), 1519143.

[13] Schaumburg, H., and Hacke, S. (2010). "Medienkompetenz und ihre Messung aus Sicht der empirischen Bildungsforschung". In *Jahrbuch Medienpädagogik 8* (pp. 147–161). VS Verlag für Sozialwissenschaften.

[14] Gapski, H. (2007). "Some reflections on digital literacy". *Proceedings of the 3rd International Workshop on Digital Literacy* (pp. 49–55). Crete, Greece: CEUR-WS.org. http://ceur-ws.org/Vol-310/paper05.pdf. [Accessed on December 22, 2020].

[15] Schüller, K., and Busch, P. (2019). *Future Skills: Ein Framework für Data Literacy*. White Paper. Hochschulforum Digitalisierung, 47.

[16] Schüller, K., Busch, P., and Hindinger, C. (2019). *Data Literacy: Ein Systematic Review*. White Paper, Hochschulforum Digitalisierung.

[17] Ganann, R., Ciliska, D., and Thomas, H. (2010). "Expediting systematic reviews: Methods and implications of rapid reviews". *Implementation Science*, 5(1), 56.

[18] Gigerenzer, G. (2015). *Risk Savvy: How to Make Good Decisions*. London: Penguin.

[19] Gigerenzer, G., Rebitschek, F. G., and Wagner, G. G. (2018). "Eine vermessene Gesellschaft braucht Transparenz". *Wirtschaftsdienst*, 98(12), 860–868.

[20] Steckelberg, A., Meyer, G., and Mühlhauser, I. (2017). "Diskussion zu dem Beitrag Gesundheitskompetenz in der Bevölkerung in Deutschland: Fragebogen nicht weiter einsetzen". *Deutsches Ärzteblatt*, 114(18), 330.

[21] Wu, M., and Adams, R. (2007). *Applying the Rasch Model to Psycho-Social Measurement: A Practical Approach*. Melbourne: Educational Measurement Solutions.

[22] Stemmler, G., Hagemann, D., Amelang, M., and Spinath, F. (2016). *Differentielle Psychologie und Persönlichkeitsforschung*. Kohlhammer Verlag.

[23] Rebitschek, F. G., Gigerenzer, G., & Wagner, G. G. (2021). "People underestimate the errors made by algorithms for credit scoring and recidivism prediction but accept even fewer errors". *Scientific Reports*, 11(1), 1–11.

Chapter 16

Mathematical Science Communication as a Strategy for Democratizing Algorithmic Governance

Florian Eyert

Weizenbaum Institute for the Networked Society,
WZB Berlin Social Science Center, Germany
florian.eyert@wzb.eu

As a consequence of digital transformation, mathematical forms of knowledge production and political processes are increasingly intertwined. On the one hand, mathematical and data-based representations of the world play a greater role in policy design and implementation — a phenomenon that can be described as algorithmic governance. On the other hand, it is becoming increasingly clear that values and political implications are often embedded in mathematical systems. This interrelation can lead to a gap between the implicit political decisions made in mathematical models and the need for their democratic legitimation, raising the question how the societal relation between mathematics and politics is to be configured in response to the resulting problems. The chapter argues that mathematical science communication can play an important role in this arrangement. However, it needs to go beyond the still common "deficit model" and explore more dialogic and participatory forms. Only if political actors and citizens are part of a broader setting of two-way communication with mathematical experts can one hope to achieve adequate democratic legitimation for practices of algorithmic governance. Some conditions and strategies for shaping such a setting are laid out and concluded with a brief illustration of the case of fairness and transparency in machine learning as well as an outlook.

Introduction

Mathematical reasoning has been a resource used in political decision-making for many centuries. But in the course of the ongoing societal processes of quantification and digitalization and the subsequent surge in available data as well as methods to analyze and act upon them, the importance of mathematical knowledge in politics has increased considerably. In such diverse areas as criminal sentencing, welfare resource allocation, policing, child abuse detection, and climate policy, applied mathematics is used to inform decision-makers or even carry out decisions automatically. In many cases these decisions are based on large data sources and inferential methods of mathematical modeling, for instance when the Swedish Public Employment Service automated decisions about individual eligibility for unemployment support or the Dutch health care system introduced a software to automatically detect children at risk of disease or abuse ([1]: 247 and 166). While such mathematical tools can have many benefits, they also have the potential to shift political agency away from citizens, place processes of societal self-governance outside the reach of democratic debate and obscure the political nature of decision-making behind seemingly neutral calculations and predictions. In short, they can create a gap between factual political consequences and the need for their democratic legitimation. Given the impact many mathematical systems used in politics have on the lives of individuals, critical reflections on the relation between mathematics and politics are needed. This chapter offers a sociological perspective on the role that mathematical science communication might play in shaping this relation. In order to address the democratic challenges posed by the role of mathematics in politics, it suggests, it is necessary to practice new forms of communication between mathematicians, politics, and the general public and to go beyond the so-called "deficit model" — according to which communication must flow from science to an ignorant society. Going beyond such one-way communication, dialogue and participation are crucial in a time of algorithmic politics. The following section prepares the argument by showing that mathematical knowledge and politics are intertwined in multiple forms and influence each other both ways. The third section then describes how, given these interrelations, different structures of communication between mathematics and politics entail different implications for the democratic embedding of mathematical knowledge in society. The fourth section aims to lay out some concrete forms mathematical science communication can take if it includes dialogic and participatory approaches, while the fifth section briefly illustrates progress and pitfalls by means of the example of fairness, accountability and transparency in machine learning. The chapter is concluded with a summary and an outlook.

Mathematics and Politics

The relation between mathematics and politics can be sketched by first describing a number of ways in which mathematical knowledge plays a role in political processes and then considering how, in turn, mathematical forms of representing the world often deviate from their alleged neutrality and can reflect, shape, and reproduce societal structures. Before doing so, it is helpful to specify an understanding of mathematics, politics, and the political that will be used.

Focusing on the practical implications of mathematical knowledge and on the role it plays in the current digital transformation, this chapter is not primarily concerned with the academic field of mathematics itself, but takes a perspective that is in one sense broader and in another sense narrower. On the one hand, it focuses on the more general area of quantitative techniques that rely on the application of mathematical concepts, including neighboring fields like computer science and data science. On the other hand, it refers only to those segments of mathematical knowledge that are regularly used in political contexts, such as statistics, machine learning and quantitative modeling, and that can also be described as "algorithmic" in the sense of computable problems. This is not to say that the described phenomena are restricted to these domains, but that they are most relevant there.

The concept of politics used here covers the institutional domain of the organization of power, including governments, bureaucracies, policies, and parliaments. The political, in contrast, will more generally be understood as anything that is related to the distribution of resources, the recognition of identities and the negotiation of power relations. It refers, therefore, to the shaping of matters of collective concern that are inherently about values and interests. Since no objectively determinable solutions exist in the political domain, this shaping must occur through societal deliberation and contestation. The political, from this perspective, is also at play in everyday life, the organization of journalistic content, public discourses, and decisions of private companies. In this sense, politics can be viewed as the organized and institutionalized practice of the political.

Mathematics in politics: Algorithmic governance

Today, mathematical methods play an important and increasingly influential role in politics. While there are various reasons for this, three can be highlighted. Firstly and most importantly, the ongoing digital transformation is affecting all areas of social life and drastically increases the availability of data as well as the demand to use them. As more and more parts of social life are moving into the digital realm, they are creating data traces that make them amenable to mathematical analysis through pattern recognition and statistical inference. The results can then be used in political debates, policy design, administrative procedures, or methods of influence, such as the microtargeting of election advertisement. Moreover, through platforms like Facebook and Twitter, the public sphere is now to a significant degree structured by the mathematical logic of content curation tools like recommender systems, and within state bureaucracies algorithmic systems promise to be efficient and cost-saving administrative tools. Secondly, the societal trend toward quantification gives increasing importance to quantitative measurements of the world, from fitness trackers to university rankings, and turns numbers into a new locus for the exercise of power [2]. Within politics, this focus on measurement manifests itself, among others, in the ideal of "evidence-based policy-making" — the notion that political decisions should be based on scientific and especially quantitative evidence — that has led to a more central role of quantitative and statistical modeling in the justification of government action. Thirdly, the growing complexity as well as political polarization that has come to shape today's societies can also be suspected to increase the use of quantitative reasoning in political negotiations, since, as Theodore Porter has shown, numbers are often perceived as a reliable and trustworthy medium of communication in situations that are characterized by heterogeneous expectations and low consensus [3].

This chapter focuses on the governance dimensions of politics. More specifically it is interested in what is beginning to be discussed as algorithmic governance: "a form of social ordering that relies on coordination between actors, is based on rules and incorporates particularly complex computer-based epistemic procedures" ([4]: 2). For the scope of this chapter, one can generally distinguish between two different ways in which mathematical knowledge might play a role in algorithmic governance: as an instrument in policy consulting, as an instrument in decision support, or even automated decision-making. The former occurs when individuals use mathematical techniques in order to develop some assessment of the world and this knowledge is then requested selectively by decision-makers to arrive at a form of governance, as it is the case in climate modeling. The latter, in turn, occurs when mathematical techniques are built into actual computational systems that are used to generate decisions repeatedly and systematically, as it is the case, for instance, when an automated system such as iBorderCtrl aims to predict a set of attributes of a person in order to allow or disallow the crossing of a border [5].

Historically, the use of mathematical knowledge in politics goes back multiple centuries. The "political arithmetic" of the 17th and 18th century was an attempt to make populations governable by making them countable in terms of births, deaths, diseases, and so on. In the early 19th century, this trend continued with an "avalanche of printed numbers" [6] and lead to the emergence of statistical knowledge, which was first thought of as a science of the state [7]. In the 20th century, the strong connection between counting populations and governing them evolved, be it through the refinement of older techniques like the census or the emergence of new paradigms like risk analysis [8] and game theory [9]. Today, mathematical knowledge arguably plays an even greater role, which the following examples aim to illustrate, standing in for many others.

Firstly, in the area of policy consulting, mathematical models of the spread of COVID-19 are an illustrative case. During the COVID-19 pandemic, governments were in the extraordinary position to require new policies directly affecting whole societies within a very short time frame and without much precedence to rely on. As a consequence, mathematical models of the pandemic played a crucial role in guiding political decisions. While "flatten the curve" became a mathematical metaphor understood by the general population and exponential growth turned into a topic of public interest, many more sophisticated models where developed in order to inform decisions between different policy options like closing schools or establishing lockdowns. These ranged from simple curve fitting to elaborate simulations in their attempts to predict the likely course of the pandemic if a given policy instruments was chosen. An example of the latter variety is the model published by Imperial College researchers in March 2020, which reportedly caused the government of the United Kingdom to initiate the first set of serious policy measures.

Secondly, algorithmic decision systems have likewise seen a spike in importance in the past years. While there are countless examples, one is standing out in terms of public and academic attention. Used in parts of the judicial system in the United States, the software COMPAS, produced by a company called Northpointe at the time, was designed to assist judges in their assessment of a defendant's likelihood to commit a crime in the future. Based on historical data it calculates a model that will assign a recidivism score to a defendant, which is then made available to judges and can play a role in determining sentences or bail conditions. Another striking example is a model used by the Public Employment Service Austria (often referred to as the AMS algorithm) to predict the chances of job seekers on the labor market and distribute resources accordingly.

The use of applied mathematics in politics has intensified not only quantitatively, but also qualitatively. While the ongoing digital transformation has brought mathematical concepts and data-driven thinking into more and more spheres of life, mathematical systems are today often not just part of political reasoning taking place in the background, but rather affect and confront individuals directly in their everyday life, be it when they interact with digital public services or with automated assessment systems. Algorithmic governance, it appears, is becoming a structural component of contemporary society.

The political in mathematics: Biases, values, and visibilities

Alongside the growing role of mathematics in politics, another phenomenon has become visible: the propensity for mathematical descriptions of the world to incorporate intended or unintended value judgments and therefore political implications. While many non-mathematicians often think of mathematics as neutral and objective, describing the real world in terms of mathematics is impossible without a number of limitations, simplifications, and background assumptions. From a sociological perspective, the notions of neutrality and objectivity appear primarily as rhetorical instruments for the legitimization of decisions by making them appear as unpolitical: "there is nearly no political decision-making process in which mathematics is not used as the 'rational' argument and the 'objective base' that is considered as replacing political judgments and power relations: mathematics as objective truth and free of politics" ([10]: 170). Scholars from the field of science and technology studies have highlighted that objectivity is a historically constructed ideal [11] and that assumptions are always inscribed in science and technology [12]. The same is true for applied mathematics, as this section aims to demonstrate. While there are countless ways in which the political can manifest in mathematical system, the illustration of some important cases can provide an intuition.

A prominent example is the issue of bias in machine learning [13]. Initially greeted emphatically and with hopes for increased efficiency and impartiality, algorithmic decision-making systems based on machine learning have been broadly criticized in recent years due to their potential discriminatory biases. A common source of such biases are data sets that omit or underrepresent certain groups or depict them in biased ways. Joy Buolamwini and Timnit Gebru, for instance, showed that facial recognition systems often discriminate on the basis of race or gender by misclassifying particular groups of people more often than others [14]. Similar biases have been found in countless other products based on machine learning, such as child abuse prediction [15] or hiring software [16]. A common effect exacerbating these risks is that of feedback loops. For example, predictive policing — the use of data analytics with the aim of sending police units to places with high crime predictions — has been shown to regularly suffer from such feedback effects. When areas with predicted crime are policed more heavily, more crimes are recorded and predictions will therefore become even higher [17]. But other forms of mathematical modeling can encapsulate particular orders of visibility and value judgments as well, as revisiting the examples of the previous section can show.

In the case of the COVID-19 models, the various available approaches each come with their particular political implications [18]. Statistical models mainly extrapolate from past trends and can lead to unjust resource allocations resulting from a status quo bias when the present or future is modeled as a mere continuation of the past. Mechanistic models based on differential equations rely on abstract categories of individuals, implicitly turning the statistical norm into a social norm and therefore failing to consider or even adversely affecting marginalized groups. More complex models like agent-based ones, too, create highly simplified depictions of individual behavior and shape political options through the selection of particular policy alternatives.

Even more strikingly, the case of COMPAS has become an often-cited example of harmful effects. As Julia Angwin *et al.* have shown, the COMPAS model negatively impacts people of color, in particular Black people, in that it systematically overestimates the risk they pose by producing more false positives in their category, which in turn leads to harsher punishments [19]. The case demonstrates that the classification of people is a highly political act (see also [20]) that can lead to significant harm if deployed maliciously or neglectfully. If mathematical classification algorithms are used for this purpose, they might increase efficiency on some level, but are often unlikely to increase neutrality. When applied to the social world, the assumptions behind mathematical techniques can become consequential and harmful for concrete individuals, especially from marginalized groups. If statistics about past behavior and social dynamics are used to predict future developments, for instance, this results in a manifest conservative and illiberal bias on the political level, as it assumes that the future will be like the past and the decisions of individuals are determined by group averages. In the case of the AMS algorithm, the partial publication of coefficients and subsequent research have made clear that it contains multiple forms of discrimination, such as an automatically lower score for women, leading to systematically different treatment by the unemployment agency [21].

These examples show that mathematical models can be political exactly in the sense that they not only fall short of representing the world in a neutral manner, but actively shape it in ways that affect lives and that are based not on abstract truth but values. From this perspective, the application of mathematical knowledge to societal problems always contains a political dimension. To the extent that algorithmic governance and therefore mathematics becomes an integral part of social and political life, there is a need for mathematics to develop reflexivity and engage in dialogue with society, as the next section will argue.

Democracy and Communication Between Mathematics, Politics and Society

The previous section has demonstrated the use of mathematical tools in governance contexts and some ways in which they themselves contain political assumptions. In an adaptation of Weingart's diagnosis of the increasing importance of scientific knowledge in social life — the *scientification of society* — and the increasing political contestation of scientific endeavors — the *politicization of science* — [22], one could assert that we are currently observing a *mathematization of society*, while the accompanying *politicization of mathematics* has yet to fully emerge. The intertwinement of two rather different fields of action — mathematical knowledge production and political negotiation and decision-making — requires reflection on the patterns of communication that exist between them in order to shape their relation in a democratic way. This section describes the general structure of important options for such communication patterns along with their limitations.

The legitimation gap between mathematics and politics

In modern societies, the relation between science and politics is complex. While scientific and mathematical knowledge production is generally conceived of as the business of scientists and mathematicians, political decision-making about matters of common concern must face the challenge of democratic legitimacy. To the extent that mathematics is used in politics, both logics become related. Peters summarizes that "[a]s experts, scientists do not possess a monopoly of relevant knowledge; values and interests will come into play" ([23]: 79). A technocratic paradigm in which the best solutions for problems are determined by scientists and then executed by governments falls short of the democratic ideal of decision-making by and for the people. As Callon *et al.* have argued, the representation of laypeople by scientists can produce similar challenges as the representation of citizens by elected politicians [24]. Considering the intertwinement sketched above, the application of mathematical knowledge from the proverbial ivory tower appears problematic, as it introduces a *legitimation gap* between mathematical knowledge production on the one hand, and societal and political decision-making on the other: The values embedded in applied mathematics can alter political outcomes without having undergone a process of democratic deliberation and legitimation. This problem is well known for political advice in general [25], but less often discussed with respect to mathematics. With the rise of algorithmic governance, however, its mathematical dimension is becoming increasingly evident. Frequently, this legitimation gap is hardly visible and becomes even more obscured when the alleged impartiality of mathematics is itself used as a means to legitimize political decisions. Moreover, in many cases the effects of algorithmic systems might even go unnoticed if those affected by it have no ability to engage with them and systematically relate their individual situations to them, which further obscures the legitimation gap. From a democratic standpoint, whenever a mathematical system is guiding a decision that would, if made by an individual, a group or an organization, be considered deeply political, this system should be assessed according to the same democratic standards. Acknowledging the important and beneficial role of science and mathematics in politics and at the same time recognizing scientific and mathematical knowledge production as a place in which political assumptions, value judgments, and particular visions of social order are embedded implies the task to find ways in which groups affected by these political effects can be involved such that unintended or harmful consequences of algorithmic governance are minimized and its premises and design are democratically shaped.

From the perspective of deliberative democracy [26], democratic legitimacy depends on the possibility of public debate and the exchange of arguments attached to different points of view. The gap between mathematics on the one hand and politics and society on the other hand constitutes a problem for this for two reasons: Firstly, the complexities of mathematical knowledge might make it prohibitive for individual laypersons to sufficiently understand the implications of systems in order to arrive at and formulate own arguments. When societal problems are translated into mathematical ones, "the group of people who could participate in the discussion of the problem and its solution becomes smaller and has a very specific composition" ([27]: 268). Secondly, due to the often assumed neutrality of mathematical knowledge and data, there are rarely institutional forms in which such exchange of arguments could take place effectively and in a democratically meaningful way. Mathematical debates about facts and political debates about values, therefore, often remain separate and algorithmic governance is outside the scope of effective democratic negotiation. It is precisely at this point that mathematical science communication can play an important role in facilitating the democratic debate needed by enabling conversations about facts *and* values.

Communication between mathematics and politics constitutes a particular case of the general problem of communication between science and politics [28]. Research in science communication, science and technology studies, the sociology of science and other disciplines has investigated how the relation between these spheres can be organized and how the communication between them shapes outcomes. The following subsections describe some structural alternatives and discuss their consequences.

Mathematical science communication and the deficit model

Historically, the default approach to science communication is the dissemination approach and its underlying *deficit model* [29]. In this view, the general public is primarily characterized by its lack of knowledge regarding a given scientific area, which is seen as the cause of limited support for science. The objective of science communication is then, primarily, to increase this knowledge through a transfer of supposedly superior expert knowledge. Science communication, in this understanding, is — implicitly or not — thought of as a one-way street: knowledge flows from science to an ignorant audience. This transfer can be done by scientists themselves or by professional science communicators, but the direction remains the same.

The deficit model is insufficient for democratically shaping the relation between mathematics and politics precisely because it does not take into account the agency of laypersons, i.e., their ability to act and voice normative positions. It models them as passive and does not account for their voices to be heard within mathematical knowledge production, not to mention for their involvement in it. Nonetheless, the deficit model is very much alive in science communication generally (see [30]) and in particular in the area of mathematics (see [31]: 35ff.), where knowledge is often considered to be especially abstract and detached from the everyday experience of citizens. Simis *et al.* argue that this persistence of the deficit model has various causes: The belief of many scientists that information is always processed rationally, the lack of communication training in the academic institutions, scientists' conception of a knowledge deficit in the public and the fact that the deficit model can easily be related to public policy [32]. As steps for going beyond the deficit model they propose to improve training for science communicators, to educate scientists in communication methods based in the social sciences and to improve the engagement of laypersons from affected communities. Against the background of the discussion so far, it appears that the first two solutions are not fully suitable for addressing many of the problems resulting from the legitimation gap. Most importantly, if the design of systems based on mathematical knowledge is seen as constant and communication and engagement are to happen afterwards, the two-way communication that enables democratic shaping of this design as well as the resulting legitimacy remain out of reach. The agency of citizens and laypersons is restricted to listening. This stresses the importance of the third solution, which is based on engagement and will be discussed further below.

Literacy

A related approach for bridging the gap between mathematics, politics, and society is the notion of literacy, which goes some of the way of addressing the agency of laypersons. Enjoying popularity in many areas today, it is often brought forward as a strategy for addressing the problems mentioned above by increasing knowledge, among the general population, among decision makers and others, and thereby helping them to become "literate" in a specific field. Applied to the areas relevant to algorithmic governance, it is discussed as mathematical literacy, statistical literacy, data literacy, big data literacy, big data infrastructure literacy, algorithmic literacy, and digital literacy. D'Ignazio and Bhargava, for instance, define data literacy as "the ability to read, work with, analyze and argue with data" ([33]: 84).

Against the background of the problem of mathematics and politics, literacy can be considered an important factor facilitating the participation in democratic deliberation about the design of mathematical systems in public life, for which a basic understanding of quantitative reasoning, statistics, data and data analysis is needed.

However, despite the importance of raising mathematical literacy, the approach has limitations as well. First of all, the problem framing of the literacy perspective is an individualistic one. It aims to address the issues resulting from the intertwinement of mathematics and politics by putting the responsibility of closing the gap between the mathematical knowledge required to understanding public systems and the knowledge that average people have about mathematics on the shoulders of the individual. This is likely to be successful for individuals that have the time, education, and resources to develop such literacy, while leaving others behind. If literacy is thought of as a strategy against opacity, i.e., the incomprehensibility of a system to an observer, at least two further major limitations can be identified. As Burrell pointed out, the limited knowledge of laypersons is only one source of opacity: it can also be caused by the inherent lack of transparency of the methods that are employed, such as artificial neural networks, or by the seclusive nature of products developed by private sector actors interested in preserving business secrets [34]. This general problem of opacity has been widely discussed in the academic literature under the "black box" concept (see [35]) and illustrates that even if mathematical literacy was fully achieved, many problems for the communication between mathematics and politics would persist because many mathematical black boxes remain opaque even to those who are "literate". Moreover, within the literature on data literacy the limitations of an understanding of literacy as merely the ability to use data has been pointed out. Sander therefore suggests a broader conception of literacy in the form of "critical big data literacy" [36] that also includes the ability to reflect on and critically assess the larger context and implications of the use of data. Nonetheless, it could be argued, the concept of literacy remains largely within the framework of the deficit model: understanding of mathematical concepts is to be achieved among a public that is not taken seriously other than as lacking precisely this understanding.

Beyond the deficit model: Dialogue and participation

While the notion of literacy takes seriously the agency of individuals in actively improving their understanding and using it, it leaves open the crucial question how a dialogue between mathematics, politics and society can be created, i.e., how laypersons voices can be heard within mathematical expert communities and how they can even participate in mathematical knowledge production despite their limited expertise. Such dialogue is especially crucial in domains in which mathematics and politics are interwoven, since without it there would be no opportunity for mathematical experts to take into account societal preferences regarding values and design decisions in a systematic way and to be responsive to critique and deliberation, i.e., to integrate conversations about facts with conversations about values in light of the factual societal diversity of such values. There is therefore a need to not only go beyond the one-way assumption of the deficit model, but also complement literacy approaches with more structural ones that take into account existing differences and inequalities. Trench has proposed to distinguish the deficit model from the dialogue model and the participation model [30]. From the perspective of democratic deliberation, both are crucial elements of a such a structural and two-way mathematical science communication.

An important precondition for dialogue is the ability and motivation of mathematical experts to take a societal perspective. To the degree that mathematical knowledge production used in political processes involves assumptions and decisions with political consequences, it can be argued that mathematicians and data scientists themselves are political actors, since they are making decisions about how the collective world is to be represented and thus acted upon [37]. They must therefore be reflexively aware of this position, actively involved in societal communication and responsive to those affected by their decisions. Going back to Simis *et al.*, one could argue that mathematicians working in applied mathematics should not only be trained in communication methods from the social sciences but also in the social sciences generally [32]. While this, of course, presents practical challenges, developing strategies for pursuing this goal would be worthwhile. Complementing the project of achieving mathematical literacy within the general population, it would require "political and social literacy" from those mathematical experts involved in building systems with societal impact. This includes an awareness of the limitations and assumptions of their own work, but also, and maybe even more importantly, a task to point out such limitations and assumptions in systems outside of their own work, where the seeming objectivity of mathematics is used as a device for legitimization. Showing shortcomings in such applications of mathematical knowledge could be a valuable task for mathematicians. But such political and social literacy is not enough on its own. To the degree that mathematical knowledge is gaining importance in public life, it becomes increasingly important for mathematicians to be engaged in and listening to the preferences and concerns of this public. This includes not only considering social questions in their work but also allowing the possibility of receiving the respective answers from those they concern. In the case of algorithmic systems this is particularly crucial, since the development of a system and the emergence of consequences for those affected by it might occur at rather distant points in time, making feedback, learning from past experiences, and engagement with criticism even more important.

Such an approach adds a second direction of listening: instead of mathematical knowledge being communicated to an "illiterate" public, experts applying mathematical knowledge are called upon to in turn listen to the concerns of those affected by them. Horst summarizes the aim of this second direction of communication succinctly: "the problem is not that publics don't listen to scientists, but that scientists don't listen to publics" ([38]: 262). In bringing both directions together, Horst proposes to think about communication between science and the public in terms of "contextual networks of negotiations over usability, credibility and influence" (*ibid.*: 263). The lesson for mathematical science communication is that depending on the context, different arrangements of communication should be considered, as the next section will explore. For any context, however, the important point is that both directions of communication are open in principle.

In the case of mathematics, such a negotiation might require a complex set of participants from different knowledge fields and societal positions in order to translate heterogeneous concepts and relate diverging concerns. Affected individuals or their respective spokespersons, mathematical experts, political decision makers, data scientists and social scientists might all be relevant in democratic negotiations about the public use of mathematical methods and models. Callon *et al.*, drawing on their characterization of the relation between scientists and laypersons as well as citizens and politicians, propose the concept of "hybrid forums" to realize such negotiations: "forums because they are open spaces where groups can come together to discuss technical options involving the collective, hybrid because the groups involved and the spokespersons claiming to represent them are heterogeneous, including experts, politicians, technicians, and laypersons who consider themselves involved [..., and] because the questions and problems taken up are addressed at different levels in a variety of domains" ([24]: 18). Such hybrid forums, as an abstract notion to be implemented in concrete institutions, are spaces where deliberation on both facts and values can occur, where different societal groups can take part in negotiations about multifaceted issues and where therefore mathematical science communication can be carried out in both a dialogic and participatory form. Moreover, hybrid forums could play a role both during the process of knowledge generation and afterwards, whereas classic science communication is often about communicating final results. For applications of mathematical knowledge to matters of societal concern, this would allow to incorporate views of stakeholders to be integrated into what could otherwise be arbitrary decisions embedded in mathematical modeling. If democratic legitimacy of algorithmic governance is to be achieved, mathematical science communication, understood broadly as societal communication about mathematics and organized around hybrid forums, presents a promising strategy for decreasing the gap that prevents it. These hybrid forums need to be actively initiated by mathematical experts as well as the organizations they are acting within and explored in a variety of forms.

Forms of Mathematical Science Communication About Algorithmic Governance

As Trench points out, "the terminology of 'dialogue' can refer to a wide range of practices and strategies" ([30]: 128). Similarly, the other structural arrangements of mathematical science communication described above are rather abstract and can be realized in a number of ways that always need to be adapted to specific contexts, as Horst has argued. Voß has stresses that, in fact, different forms of engaging the public in science and technology imply different forms of political organization and representation [39]. This section aims to give a brief overview of some relevant dimensions along which efforts to practice mathematical science communication about algorithmic governance can vary if a broad and two-way understanding of science communication is applied. Throughout it should be kept in mind that — from a sociological standpoint — all of these approaches have their limitations and always only enable at most a partial opening of mathematics to society (see also [40]).

Levels: Individual, organizational and society-wide

Firstly, mathematical science communication can occur at different levels. Often, it is located on the level of an individual person engaging in mathematical science communication using a variety of channels, such as news media, social media, popular science publications, podcasts, and public forums. Beyond this, it can also be thought of on a larger scale as the effort and responsibility of organizations, which could be science communication divisions of universities, but also digital platform companies, NGOs, and others. Rödder points out that "organizations shape the image of science in society to a degree hitherto underexplored by science communication research" ([41]: 179). Especially in cases where resources, abilities, or responsibilities exceeding those of an individual are needed to establish successful dialogue and participation or knowledge is strongly linked to organizational logics, as it is the case with algorithmic systems kept secret for business reasons, organizations can be an important unit of mathematical science communication. Lastly, there could be a society-wide deliberation in the form of public debates or civil society activities in which larger parts of society are engaged. This would be relevant for issues that are perceived as critical enough to affect all of society in a significant way. In these cases, mathematicians themselves could act as part of civil society and social movements. A good example of such a society-wide communication effort that contributed to the "politicization of mathematics" are the protests against high school grade calculations that took place in the UK in 2020. After the OFQUAL agency had used a simple statistical model to predict grades for the students who could not participate in the final exams due to the pandemic, students and scholars alike began to criticize flaws in the assumptions of the model, such as using the grades achieved at a school in the past to determine the grades of individual students of this school in the present. Eventually, they effected a recalculation of the grades based on assessments by teachers rather than the algorithmically determined predictions.

Involvement: Actors and formats

Another dimension relates to who is actually involved in societal communication about mathematics and how this involvement can be structured. On the so-called expert side, mathematicians, data science practitioners, professional science communicators and science journalists are some of the actors that could take part in mathematical science communication. On the so-called laypersons side, the general public thought of as a homogeneous entity is just one participant of communication besides various sub-publics, societal groups, stakeholder organizations, individual citizens, civil society actors, and politicians. Interdisciplinary communication with other fields of knowledge can be a fruitful contribution to mathematical science communication as well, especially in the case of the social sciences and humanities. If mathematical science communication is to take a dialogic and participatory form, it needs to go beyond the model in which communication divisions in universities and research centers simply communicate the results of science, and instead make the concrete implementation of the theoretical concept of hybrid forums a relevant concern. Among the possible formats for such forums are consensus conferences aiming to achieve exchange between heterogeneous stakeholders, although they are sometimes accompanied with inflated expectations that lead to disappointment (see [38]). Another avenue consists in the notion of participatory modeling. Here the relevant stakeholders are brought into the process of creating mathematical models itself in order to establish feedback during development. Participatory modeling can be used, among many others, in simulation models [42] and matching algorithms [43]. It does, however, pose a risk of "participation-washing" [44], in which participation methods known to not be particularly effective are used as an instrument of legitimization. Given such risks of ad hoc participation initiatives, more institutional and structural arrangements for implementing hybrid forums, be it through publicly funded organizations or civil society actors, are necessary as well. Lastly, the institutionalized reaction to and incorporation of uninvited participation [45], such as protests, can contribute to an inclusive negotiation of the role of mathematical methods and models in society. In all cases, experimentation with forms of involvement is crucial.

Abstraction: From precise to general

Furthermore, mathematical knowledge can be communicated on different levels of abstraction, with each having advantages in different contexts. At one end of the spectrum, mathematical science communication can be organized around an understanding of particular aspects of mathematics in a precise way, i.e., on a level of granularity that would be accepted in academic mathematics itself. At the other end of the spectrum, focusing only on the general assumptions and strategies used in particular mathematical methods can be of value as well. Especially in the current context, where machine learning is widely used but rarely understood by the general public, the more abstract and general approach can often be appropriate. In this case, it is not precise details about parameter optimization or model selection that are relevant for democratic debate, but rather the general premises of data-driven decision-making, such as the assumptions and limitations behind constructing predictions from past events and incomplete samples in the context of complex social systems. Here, mathematical science communication could engage in debates about general methodological characteristics and flaws in applications of mathematical knowledge, bringing mathematical reasoning within the radar of democratic deliberation, offering sober explanations and making mathematical methods debatable without advanced expertise.

Scope: From self-referentiality to intervention

Science communication about mathematics can, lastly, be situated on a spectrum from self-referentiality to intervention. In the case of self-referentiality, what is communicated relates primarily to the scientific work of the communicator her- or himself or is communicated in place of the creators of the communicated knowledge. This can mean communicating research results to the broader public or engaging in dialogue about the premises of one's own work. On the other hand, mathematical science communication can take the form of interventions into projects outside of the scope of the communicator's work. This could happen by pointing out the implications of the use of a given mathematical tool in the public sector, criticizing assumptions made, or generally bridging the cognitive gap between those building a given system and those affected by it. In the current situation, where "artificial intelligence" is on the height of its hype cycle, mathematicians can engage in science communication not just in order to educate the public about the functioning of such systems but also to discuss their limitations. Breaking hype waves instead of riding on them would be a meaningful way in which mathematicians could contribute to the preconditions of deliberative and inclusive debate about the introduction of new technologies like AI. Mathematicians can then actually engage in critique and challenge assumptions about efficiency and neutrality in those cases where they cannot be justified from a purely mathematical standpoint. They would assume the role of "public intellectuals" (see [10]: 182) in the sense that they use their expertise in order to engage in matters of public concern. An example of this is the work of Cathy O'Neil, who was one of the first mathematicians to publicly and effectively criticize the implications of algorithmic systems in a way that made mathematical concepts understandable to a broader audience [46]. Contributing to "critical mathematical literacy" in this way could likewise be a task for science journalism to take on ([31]: 46f.). While the interventionist form of mathematical science communication is not usually considered part of the standard, it is becoming more and more urgent given the current spread of algorithmic governance into many parts of civic life.

The Case of Fairness, Accountability and Transparency in Machine Learning

The centrality of machine learning in many of the systems discussed above requires taking a brief look at a particular response to some of its problems as an illustrative example. As laid out so far, machine learning is becoming an important part of many societal processes through its application in systems for automated or at least semi-automated decision-making in a variety of domains, such as welfare, criminal justice, border control and media selection. In many cases, concerns with injustice and lack of democratic oversight have arisen from them, often relating to data sets biased against particular populations, inappropriate generalizations, opaque "black box" systems and a lack of opportunities for individuals to intervene or receive explanations. In response to these problems, a movement in academia and the technology industry has begun to address them under the label of "Fairness, Accountability and Transparency in Machine Learning" (FAccT), with a similarly named yearly conference at its center. In this context, technical instruments have been developed to automatically detect and correct biases in machine learning models that are known to contribute to the reproduction or intensification of inequalities. This is usually done by defining a fairness metric — of which there are many [47] — and then potentially adjusting data sets or classifiers to these metrics. Furthermore, there are attempts to increase the transparency of machine learning systems through the use of "Explainable AI" that is able to specify general reasons for a given classification to its user, for instance when highlighting relevant regions in an image classification.

From the point of view of this chapter, these issues and responses to them can be interpreted against the background of mathematical science communication, as they are precisely about the embedding of mathematical methods and models in societal processes and the mitigation of resulting democratic challenges. Initially, FAccT set out as a largely technical endeavor that attempted to incorporate general demands for fairness and transparency into mathematical frameworks. While this meant acknowledging non-mathematical and non-technical concerns through the development of social and political literacy, it was only of limited success with regard to engaging different societal groups into actually formulating and contributing their claims for equality and participation. Parts of the research field overemphasized the capacity of mathematical formalizations to function as practically useful encapsulations of complex notions like justice, which are in fact always historically situated and negotiated. The multiplicity of existing fairness metrics and the impossibility to decide between them on the grounds of mathematics alone demonstrates the importance of inclusive conversations about both facts and values. This requires actual dialogue with and participation of stakeholders in the development of machine learning systems and cannot be circumvented by anticipating their result in generalized mathematical definitions of fair systems that do not take into account the social conditions and the concrete concerns and demands of affected groups. Recently, the field has become more and more diverse, opening up to social sciences, legal scholars, and the humanities. Increasingly it is taking the form of a hybrid forum through the establishment of spaces of interdisciplinary communication about the social context of mathematical knowledge production. Its growing impact and the visibility of the debates emerging from it show that through interdisciplinary collaboration and the active consideration of societal values in the design of mathematical systems it is possible to challenge current forms of algorithmic governance and point to strategies for its democratization. Nonetheless, FAccT remains a largely academic project and the inclusion of laypersons and political decision-makers is still a challenge. From the perspective of mathematical science communication, the field of FAccT could therefore benefit from further engaging actors outside of academia, collaborating with them in the sense of hybrid forums and seeking public intervention wherever possible. A contributing factor to the success of such hybrid forums might also be the level of public salience of the relevant issues. The recent documentary "Coded Bias", which explains many of the FAccT topics in an approachable manner and created broader public awareness for the issue of bias in machine learning, could in this sense be seen as one element of the larger project of mathematical science communication.

Conclusion

This chapter has demonstrated that there is an increasing connection between mathematical forms of knowledge and political processes, intensified through the ongoing digital transformation and apparent in the phenomenon of algorithmic governance, and that this connection tends to produce an often invisible legitimation gap between mathematical systems and democratic requirements. It has argued that an important task for mathematical science communication is to decrease exactly this gap by engaging in and facilitating participation and dialogue between the field of mathematics and societal stakeholders. This can take place in various arenas and on various levels, it can include various kinds of actors and many different formats, each of which has advantages and limitations. The case of FAccT illustrates that dialogue is necessary and societal claims cannot simply be incorporated into mathematical formalizations, because the negotiation of collective meaning, which is the foundation of deliberative democracy, cannot take place within mathematics itself. Critical and multidisciplinary research can, however, provide an important point of departure for such dialogue.

For the future, a number of overarching challenges can be identified. Firstly, as digital transformation progresses it can only be expected that mathematical methods and models will play an even deeper and more important role in our lives, intensifying the accompanying risks along the way. Secondly, the increasing specialization of mathematical methods for data analysis are likely to make it equally more difficult for citizens to engage in and critically reflect upon the effects these methods might have on them. Thirdly, with the continued reconfiguration of the relation between the commercial and the political sphere through private providers of public digital services as well as digital platform companies fulfilling public functions, the opacity of many democratic infrastructures is likely to become an even more important issue. All of these developments call for a vigorous expansion of the efforts of mathematical science communication, understood broadly and as one tool among others to strengthen the democratic fabric of contemporary societies.

References

[1] AlgorithmWatch (2020). "Automating Society". In Fabio Chiusi, Sarah Fischer, Nicolas Kayser-Bril, and Matthias Spielkamp (eds.). https://automatingsociety.algorithmwatch.org/.

[2] Mau, S. (2019). *The Metric Society: On the Quantification of the Social*. Cambridge, UK: Polity Press.

[3] Porter, T. M. (1996). *Trust in Numbers: The Pursuit of Objectivity in Science and Public Life*. Princeton: Princeton University Press.

[4] Katzenbach, C., and Ulbricht, L. (2019). "Algorithmic governance". *Internet Policy Review*, 8(4), 1–18.

[5] Sánchez-Monedero, J., and Dencik, L. (2020). "The politics of deceptive borders: 'Biomarkers of Deceit' and the case of iBorderCtrl". *Information, Communication & Society*, 25(3), 413–430.

[6] Hacking, I. (1982). "Biopower and the avalanche of printed numbers". *Humanities in Society*, 5, 279–295.

[7] Porter, T. M. (1988). *The Rise of Statistical Thinking, 1820–1900*. Princeton: Princeton University Press.

[8] Bouk, D. (2018). *How Our Days Became Numbered: Risk and the Rise of the Statistical Individual*. Chicago: Universiy of Chicago Press.

[9] Hagemann, H., Kufenko, V., and Raskov, D. (2016). "Game theory modeling for the cold war on both sides of the iron curtain". *History of the Human Sciences*, 29(4), 99–124.

[10] Keitel, C., and Vithal, R. (2008). "Mathematical power as political power — The politics of mathematics education". In Philip Clarkson and Norma Presmeg (eds.), *Critical Issues in Mathematics Education* (pp. 167–188). Boston, MA: Springer US.

[11] Daston, L., and Galison, P. (2007). *Objectivity*. New York: Zone Books.

[12] Akrich, M. (1992). "The de-scription of technical objects". In Wiebe E. Bijker and John Law (eds.), *Shaping Technology/Building Society* (pp. 205–224). Cambridge: MIT Press.

[13] Barocas, S., and Selbst, A. D. (2016). "Big data's disparate impact". *California Law Review*, 104(3), 671–732.

[14] Buolamwini, J., and Gebru, T. (2018). "Gender shades: Intersectional accuracy disparities in commercial gender classification". *Proceedings of Machine Learning Research*, 81, 1–15.

[15] Chouldechova, A. (2017). "Fair prediction with disparate impact: A study of bias in recidivism prediction instruments". *Big Data*, 5(2), 153–163.

[16] Raghavan, M., Barocas, S., Kleinberg, J., and Levy, K. (2020). "Mitigating bias in algorithmic hiring: Evaluating claims and practices". In *Proceedings of the 2020 Conference on Fairness, Accountability, and Transparency* (pp. 469–481). Barcelona, Spain: ACM.

[17] Ensign, D., Friedler, S. A., Neville, S., Scheidegger, C., and Venkatasubramanian, S. (2018). "Runaway feedback loops in predictive policing". *Proceedings of Machine Learning Research*, 81, 1–12.

[18] Eyert, F. (2020). "Epidemie und Modellierung. Das Mathematische ist politisch". *WZB-Mitteilungen*, 168, 82–85.

[19] Angwin, J., Jeff, L., Surya, M., and Lauren, K. (2016). "Machine bias". In *ProPublica*. https://www.propublica.org/article/machine-bias-risk-assessments-in-criminal-sentencing.

[20] Bowker, G. C., and Star, S. L. (2000). *Sorting Things Out: Classification and Its Consequences*. Cambridge, MA: MIT Press.

[21] Lopez, P. (2019). "Reinforcing intersectional inequality via the AMS algorithm in Austria". In *Conference Proceedings of the STS Graz Conference 2019. Critical Issues in Science, Technology, and Society Studies* (pp. 289–309).

[22] Weingart, P. (1983). "Verwissenschaftlichung der Gesellschaft — Politisierung der Wissenschaft". *Zeitschrift für Soziologie*, 12(3), 224–241.

[23] Peters, H. P. (2014). "Scientists as public experts: Expectations and responsibilities". In Massimiano Bucchi and Brian Trench (eds.), *Routledge Handbook of Public Communication of Science and Technology* (pp. 70–82). London: Routledge.

[24] Callon, M., Lascoumes, P., and Barthe, Y. (2009). *Acting in an Uncertain World: An Essay on Technical Democracy*. Cambridge, MA: MIT Press.

[25] Lentsch, J., and Weingart, P. (2011). *The Politics of Scientific Advice. Institutional Design for Quality Assurance*. Cambridge, UK: Cambridge University Press.

[26] Habermas, J. (1996). *Between Facts and Norms: Contributions to a Discourse Theory of Law and Democracy*. Cambridge, MA: MIT Press.

[27] Sánchez Aguilar, M., and Blomhøj, M. (2016). "The role of mathematics in politics as an issue for mathematics teaching". In Paul Ernest, Bharath Sriraman, and Nuala Ernest (eds.), *Critical Mathematics Education. Theory, Praxis and Reality* (pp. 253–271). Charlotte, NC: Information Age Publishing.

[28] Renn, O. (2017). "Kommunikation zwischen Wissenschaft und Politik". In Heinz Bonfadelli, Birte Fähnrich, Corinna Lüthje, Jutta Milde, Markus Rhomberg, and Mike S. Schäfer (eds.), *Forschungsfeld Wissenschaftskommunikation* (pp. 183–205). Wiesbaden: Springer.

[29] Wynne, B. (1991). "Knowledges in context". *Science, Technology, & Human Values*, 16(1), 111–121.

[30] Trench, B. (2008). "Towards an analytical framework of science communication models". In Donghong Cheng, Michel Claessens, Toss Gascoigne, Jenni Metcalfe, Bernard Schiele, and Shunke Shi (eds.), *Communicating Science in Social Contexts* (pp. 119–135). Dordrecht: Springer Netherlands.

[31] Hartkopf, A. M. (2020). "Mathematical science communication. A study and a case study", PhD thesis. https://refubium.fu-berlin.de/handle/fub188/28703.

[32] Simis, M. J., Madden, H., Cacciatore, M. A., and Yeo, S. K. (2016). "The lure of rationality: Why does the deficit model persist in science communication?". *Public Understanding of Science*, 25(4), 400–414.

[33] D'Ignazio, C., and Bhargava, R. (2016). "DataBasic: Design principles, tools and activities for data literacy learners". *The Journal of Community Informatics*, 12(3), 83–107.

[34] Burrell, J. (2016). "How the machine 'thinks': Understanding opacity in machine learning algorithms". *Big Data & Society*, 3(1), 1–12.

[35] Pasquale, F. (2015). *The Black Box Society: The Secret Algorithms that Control Money and Information*. Cambridge: Harvard University Press.

[36] Sander, I. (2020). "What is critical big data literacy and how can it be implemented?". *Internet Policy Review*, 9(2), 1–22.

[37] Green, B. (2018). "Data science as political action: Grounding data science in a politics of justice". https://arxiv.org/abs/1811.03435.

[38] Horst, M. (2008). "In search of dialogue: Staging science communication in consensus conferences". In Donghong Cheng, Michel Claessens, Toss Gascoigne, Jenni Metcalfe, Bernard Schiele, and Shunke Shi (eds.), *Communicating Science in Social Contexts* (pp. 259–274). Dordrecht: Springer Netherlands.

[39] Voß, J.-P. (2019). "Re-making the modern constitution: the case for an observatory on public engagement practices". In Dagmar Simon, Stefan Kuhlmann, Julia Stamm, and Weert Canzler (eds.), *Handbook on Science and Public Policy* (pp. 67–90). Cheltenham: Edward Elgar Publishing.

[40] Dickel, S., and Franzen, M. (2015). "Digitale Inklusion: Zur sozialen Öffnung des Wissenschaftssystems". *Zeitschrift für Soziologie*, 44(5), 330–347.

[41] Rödder, S. (2020). "Organisation matters: Towards an organisational sociology of science communication". *Journal of Communication Management*, 24(3), 169–188.

[42] Halbe, J., Holtz, G., and Ruutu, S. (2020). "Participatory modeling for transition governance: Linking methods to process phases". *Environmental Innovation and Societal Transitions*, 35, 60–76.

[43] Lee, M. K., Psomas, A., Procaccia, A. D., Kusbit, D., Kahng, A., Kim, J. T., Yuan, X., *et al.* (2019). "WeBuildAI: Participatory framework for algorithmic governance". *Proceedings of the ACM on Human-Computer Interaction*, 3, 1–35.

[44] Sloane, M., Moss, E., Awomolo, O., and Forlano, L. (2020). "Participation is not a design fix for machine learning". https://arxiv.org/abs/2007.02423.

[45] Wynne, B. (2007). "Public participation in science and technology: Performing and obscuring a political–conceptual category mistake". *East Asian Science, Technology and Society: An International Journal*, 1(1), 99–110.

[46] O'Neil, C. (2016). *Weapons of Math Destruction: How Big Data Increases Inequality and Threatens Democracy.* New York: Crown.

[47] Friedler, S. A., Scheidegger, C., Venkatasubramanian, S., Choudhary, S., Hamilton, E. P., and Roth, D. (2019). "A comparative study of fairness-enhancing interventions in machine learning". In *Proceedings of the Conference on Fairness, Accountability, and Transparency* (pp. 329–338). Atlanta, GA: ACM.

https://doi.org/10.1142/9789811253072_0018

Chapter 17

Articles About Mathematics, Mathematics Behind Articles

Andreas Loos

ZEIT ONLINE, Germany
andreas.loos@zeit.de

Mathematics (more generally: mathematical sciences) and publishers of journalistic content in general coexist in a complex alliance. We will focus here on two aspects of this relationship: on the one hand, journalists are deriving news stories from actual scientific results or events in the mathematical community; they are producing *articles about mathematics*. On the other hand, journalists can benefit from mathematical research by applying maths itself in the process of producing the news stories; they are producing news *by the means of mathematics*.

Introduction

Mathematical research and mathematical work is complex and requires focus and concentration. It is an exact science and it is based on the principle of mathematical proofs. It is also an auxiliary science for numerous applications in quantitative sciences, including engineering and natural sciences, as well as, for example, for quantitative social research. However, in the public perception, mathematics is usually strongly identified and overlaid with the contents of the school subject with the same name, which in truth is only a distant relative of contemporary mathematics done nowadays in research institutions. These and other characteristics and facets of mathematics are reflected in typical public images of mathematics (see e.g., [1,2]), and these images sometimes appear to be heavily distorted from the point of view of research mathematicians.

We cannot describe in detail the emergence of these images, although they provide the background for the role which mathematics plays in and for many media publishers, and in particular many online news sites (which we will be focusing on in the following sections). For publishers of journalistic content, mathematics can play in many ways a role of growing importance, both internally and externally. In this work, we will deal with two aspects:

1. Mathematical content (for instance the work of a mathematician or a research result) can be transformed directly into journalistic content for a broad audience.

2. Mathematics can also operate indirectly as a catalyst enabling journalists to cover or analyze a subject (not necessarily a mathematical one, as we will see).

Mathematics as Subject

A typical example of concrete news with a basis in mathematical content is articles on mathematically gifted persons — a subject which obviously has a certain attributed news value. (For news factors and values see for example, [3,4,19], and Figure 1. It should be noted that the concept of news factors has been developed in connection with news, but in principle also applies to any journalistic articles in general.)

If mathematics is the reason or core content of a publication for a broad audience, then this fact is almost always revealed to the recipient. It is usually explicitly stated that this is an article about mathematics (which does not imply that the mathematics is explained, as we will see).

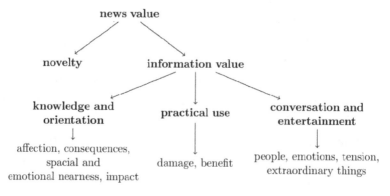

Figure 1: New factors according to Schwiesau and Ohler [4].

The form in which the mathematical aspect is presented, however, depends on many factors, for example, on the subject and the news factors attributed to the subject. It also differs with the respective medium as well as the narrative approach chosen by the journalist. The latter is even subject to changes over time, depending on, amongst other things, the role and self-perception of the journalist. For example, science journalists in the 1980s often saw themselves as mediators and explainers of knowledge (good examples are the "Knoff-Hoff-Show" and the "Wissenschaftsshow" in German TV). In the 1990s and around the turn of the millennium, science journalism shifted the focus to content that was supposed to have a practical use in everyday life. Currently, reporting and "checking" of facts are considered very important. In the fake news debate, there is a growing need for security, and there is some hope that this need can be satisfied by mathematically approved insights and figures.

Parts of mathematics can in fact connect to these trends and, moreover, to a selection of news factors. On the other hand, it is obvious that a major part of mathematical research is neither applicable to the everyday lives of the users nor can it ever be communicated to a broad audience on a simple level. This is a (sometimes frustrating) challenge for those scientists who would like to show the science in as many aspects as possible to a broad public.

The majority of journalistic texts in mass media either deals with social/political topics or they are intended to have a certain use for the users by giving an advice or providing some other service — or they simply are supposed to entertain the users. All in all, mathematical topics have a very low total attributed news value, with the exception of mathematics in education, training or career.

To support this thesis, we have counted and analyzed the articles related to mathematics published on faz.net, spiegel.de and zeit.de in 2019 (to exclude COVID-19 as a dominating subject from 2020 on). In the article archives of the respective newspapers, we searched for texts containing the word "Mathematik" (mathematics, mathematician) either in the text body or in the header (e.g., teaser or headlines). Picture galleries, videos, and audios were excluded from the search, as well as texts containing mathematical topics in a broader sense, but not the keyword "Mathematik" itself (e.g., articles on statistics, computer science, artificial intelligence, algorithms, etc.).

In total for 2019, we found 175 articles at spiegel.de, 183 articles at faz. net and 86 articles at zeit.de. (We note that each of these media published tens of thousands articles in the same time span.) These articles can be separated into three groups:

- articles with a clear mathematical reference;
- articles referring to education and career, with mathematics as a side aspect; and
- articles containing the word "mathematics", but having nothing or very little to do with mathematics.

The third group made up the largest share (zeit.de: 58%, faz.net: 41%, spiegel.de: 47%). In these articles, mathematics was often used as a metaphor, as in the following example:

> "Man muss ja nicht Mathematik studiert haben, um zu begreifen: Es sind nach wie vor elf Spieler auf dem Platz zu verteilen, und zieht man einen Spieler von vorne nach hinten, fehlt er halt vorne." ("You do not need to have a degree in mathematics to understand: One still has to put eleven players on the [football] field and if you move one player from the front to the back, then he is missing in the front.") [5]

Other examples were a book review mentioning that author Eugen Ruge once studied mathematics and an article about population growth posing questions to a future scientist (and mathematician) predicting a shrinking world population because of social (!) reasons.

The second largest share of articles (zeit.de: 33%, faz.net: 44%, spiegel.de: 36%) was dealing with mathematics as a school subject or as a career and educational issue, in particular in connection with the final secondary school examinations in mathematics and their difficulty level. This issue was heavily debated in 2019 in many federal states in Germany.

Only very few articles dealt with mathematics or mathematical research as a core subject or "hook". Among these, we found the weekly mathematical puzzle on spiegel.de. (Only a small selection of these puzzles were actually in the set of articles, since most of them do not contain the word "mathematics"). After removing these puzzles, we got 12 articles at spiegel.de (7% of all articles with the keyword mathematics), 15 articles at faz.net (8%) and four articles at zeit.de (5%).

Of course, the topics of these articles are overlapping, in part due to the current news situation (for instance: important awards, discussion of synchronized timetables for trains in Germany). In detail, we found the following topics:

- the Abel Prize being awarded to mathematician Karen Uhlenbeck and the Zuse Medal to mathematician Dorothea Wagner;
- portraits of a Sudoku and a poker master with winning or solving strategies for these games;
- portraits of a mathematically gifted child with autism, a highly gifted pupil with a special interest in mathematics, a highly gifted student of mathematics (who is an NFL athlete with Ph. D. in mathematics) and a woman who can recite more than 10,000 digits of π by heart;
- portraits of Alan Turing, René Descartes, Emil Julius Gumbel (a mathematician, who statistically analyzed right-wing tendencies of the Weimar judiciary), France Poppy Northcutt (a software developer at NASA) and Vladimir Vojewodski (a contemporary mathematician);
- the network of persons around the proof of the Poincaré conjecture by Grigorij Perelman;
- an interview with the Bonn mathematician Michael Rapoport about his mathematical research;
- an interview with a statistician about prediction markets;
- the development of a new computer chip for doing computations in non-Euclidean geometries;
- the simulation and visualization of a computing machine with marbles (Youtube);
- how the date for Easter is determined each year;
- articles on delays and the optimization of the schedule at *Deutsche Bahn*;

- an article on crocheting geometric forms;
- the naming of "Napoleon's Theorem";
- a review of the book *99 Variations on a Proof* by Philip Ording;
- the integral solution of the equation $x^3 + y^3 + z^3 = 42$ by the British mathematician Andrew Brooker;
- the problems and difficulty of the Mathematical Olympics 2019; and
- and a new algorithm to find parking sites, developed by US physicists Paul Krapivsky and Sidney Redner.

This choice of topics shows clearly the news factors and values that are commonly assigned to mathematics. For mathematics in the educational department (in particular the articles about the secondary school examinations), this is obvious (affection, consequences, damage and benefit, emotions etc.). For the other mathematical topics, the following seem to hold:

Knowledge and Orientation Results from mathematical research like the solution of the weak Goldbach Conjecture or the classification of finite simple groups usually have only a small influence on everyday life. Users are typically not (directly) affected by new results, and if so, then it is not easy to tell in which way. However, if something has an impact inside the mathematical community and has an emotional aspect, it can tunnel outside: Recipients do not need to understand what geometric partial differential equations and gauge theory actually are, when they are told that some woman (the first, by the way) got the Abel Price. The content of the Poincaré conjecture is not as "interesting" as quarrels about its solution and the fights between researchers.

Practical Use Mathematicians can be experts who can analyze and comment developments. The interview with an optimization expert for the train system and the analysis of the delays at *Deutsche Bahn* fall into this category.

Conversation and Entertainment Most of the articles on mathematics are supposed to be entertaining and carrying some kind of *party* *knowledge*, such as Puzzles, the Mathematical Olympics, the solved problem about the number 42, crocheting mathematics, the strategy for finding a parking site and the Easter formula. (The latter two are of limited practical use and merely for fun.) Even the many texts on gifted people can be categorized here (apart from the emotional closeness that is also created).

To summarize: The articles on mathematics we found, are dealing, to a large extent, with news factors in the entertainment section. Mathematics is presented as a "freaky" science and often placed in the context of autism, highly talented people, and "nerds". If an article contains mathematics more concretely, it does not go beyond the school level (basic arithmetic, compass, and ruler); talking about proofs is a taboo. (For example, the book review of *99 Variations on a Proof* does not even mention the statement that is proven 99 times in the book.) Sometimes, mathematicians are also seen as experts or are invited to act as advisors on social issues, for example when they are talking about how to solve distribution problems. In this context one can often see mathematics in kind of a service function, mathematics is presented as a tool to cope with problems.

Mathematics as a Catalyst

It is a more recent phenomenon to actually *use* mathematics in the journalistic work process. This goes hand in hand with the emergence of data journalism and the entering of data scientists or people with a mathematical background into newsrooms. Data journalism explores and evaluates data sources (like statistical data or measurements) and generates journalistic findings from that; these findings are communicated to recipients by presenting the contents appropriately.

The evaluation of data and results and the communication of the findings are often based on the application of methods from mathematics or computer science; one could think here, for example, of the use of cluster algorithms in the visualization of data. In contrast to above, the mathematics remains behind the scenes and is usually not explained to the recipients (or only very briefly) since it is not the core topic of the text; without mathematics, however, the evaluation would not be possible in this form and depth. Nevertheless, here is an opportunity for computer science and scientific statistics to make use of new results from research (algorithms, ideas) and to bring them into newsrooms. Below we will briefly outline in which aspects one can speak of scientific work being carried out in journalism.

In connection with data journalism, it is interesting that mathematics can, in particular, act as an exact and auxiliary science for reliable conclusions from data evaluations. This fact feeds a narrative which is often and mistakenly read in the reverse direction: the application of mathematics is supposed to extract truth even from bad data. The wrong belief is that if only bad, noisy, or not enough data is available, then a mystery "application" of mathematics can improve or qualify the data — even if it remains unclear how this should work in practice. This wish for "math-magical blessing" of data is not uncommon, and it is an often overlooked ethical problem for mathematicians, who have to decide how far to go when extracting information from data.

This problem arises, for example, when a provider of more or less anonymous online surveys promotes the purported quality of its "representative" survey results by means of a many-page-paperwork full of mathematical formulae [17]. Even science itself is not exempt from the fact that mathematical methods can give and take away meaningfulness to and from measured data; think of the long-simmering debate on significance and the p-value [6]. Data journalism also has to deal critically with this phenomenon.

Data journalism

Data journalism is a young and diverse field of work. Historically it is often placed in the tradition of the older *computer-assisted reporting* and the younger so-called *precision journalism*, which was developed as a contrast to *new journalism* in the 1990s. One of the core ideas of new journalism was to adapt means from literature and creative writing to journalism. Precision journalism, in contrast, tried to apply methods from social sciences to journalism [7].

Even though the first approaches for data-based journalism already existed in the 19th century [8], data journalism in the modern sense did not develop until the turn of the millennium, parallel to online journalism. The latter enabled interactive visualizations, an essential form of presentation in data journalism. In [9], the form of presentation is counted among the four core aspects of data journalism. According to Weinacht and Spiller, data journalism is:

1. a form of research that "wants to read stories from records";
2. a form of (statistical) interpretation of journalistic findings;
3. a form of presentation (graphic, interactive); and/or
4. journalism that publishes the raw data set.

Big hopes are connected with these aspects, including:

1. more objectivity, credibility, and security (for example by evaluating larger amounts of data);
2. more citizen participation/open access to data sources ("data literacy"), especially if the medium also allows the recipients to explore the data themselves; and
3. an individualization/personalization of the presentation, in particular through new forms of visual processing.

It should be noted here that there is a critical discussion about the extent to which data journalism can in fact achieve these goals, see for example [2,10]. The key point is that journalism is driven by telling the truth and about being as objective as possible, but at the same time it is about selecting and weighting information (and therefore highly subjective). So the argument is on the one hand that more data may provide a better approximation of reality — on the other hand, more data is simply more information to choose from, and the degree of objectivity is at the end determined by what is chosen. It is also not clear if presenting (sometimes lots of) data to users really enhances understanding, and, moreover, if users really do want to explore data by themselves. However, it may be that data literacy arises with time, and that we are at the moment in a learning process, and users are also learning to deal with complicated data dashboards on news sites.

A nice example of current data journalism and the use of statistics in journalism is the project FiveThirtyEight by journalist and blogger Nate

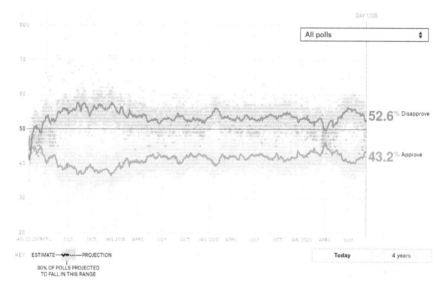

Figure 2: Since 2007/2008, the project FiveThirtyEight is analyzing and weighting more than 450 regular opinion polls to compute a projection of the popularity of Democrats and Republicans with statistical methods. The aim is to predict (as good as possible) who will win the next presidential election.

Silver [11] (Figure 2), a more recent example for excellent data journalism is the analysis of excess deaths due to COVID-19 in *The Economist* [12]. The latter clearly shows the direction in which data journalism is developing at the high end: the clever and skillful application of very intricate tools from statistics and machine learning to extract stories and facts from (even incomplete) data.

However, at the other side of the spectrum, you still find that simple counting of values in some tables to generate bar charts is often also called "data journalism".

Scientific work

Weinacht and Spiller also criticize data journalism: "In data journalism there is often a mountain of data at the beginning and any research question becomes a superfluous trivial matter when only spectacular information is found in the data" [13].

Indeed, scientists and journalists have different goals. Scientific research is not the (main) task of journalists, since they are supposed to produce journalistic content. Scientists are developing theory and the system of science, while journalists are fulfilling a democratic and social task and are satisfying the interests of the recipients (e.g., political education, explanation and discussion of social problems, entertainment).

And there are even more differences, if you look at data journalism in particular. In the quantitative sciences, one generates hypotheses. Data is subsequently collected to support or falsify them. In contrast to this, the object of research for data journalists — if you like — *is* the data which has been made available to them. Naturally, their hypotheses can therefore only be generated *after* data has been generated. In this respect, however, data journalists do indeed (sometimes) follow a scientific approach: like scientists, their first aim is to gain information and knowledge from their object of research. In short: Scientists search for patterns, structures, and questions in reality by means of data. In contrast to this, the work of data journalists is in general "data-driven", searching for patterns, structures (and questions or holes) in the data to reflect reality.

However, there is a long list of touching points and parallels between science and data journalism, containing the following:

Formulating Hypotheses As we will see in the examples below, in data journalism indeed hypotheses on data sets are regularly formulated, discussed and checked against the data.

Scientific Comparison Obtaining expert knowledge and even reading scientific literature to interpret one's own findings is considered part of good journalistic research.

Critical Handling of Data Sometimes critical data analysis in the newsroom can reveal errors in scientific studies based on the same data. Neither science nor journalism is immune to errors in or misunderstandings of data.

Software and Technology For a long time, research institutions and high-tech enterprises have had a leading position in technology. However, their advance is shrinking. Programming is no longer uncommon in newsrooms and open-source software libraries are being used in the same way as in science. Science and journalism are converging in this aspect.

Methods and Algorithms In journalism, numerous algorithms, especially from computer science, can be applied, especially for clustering, calculating decision trees/random forests and for regressions.

Many newsrooms are quite capable, in terms of staff and methods, of using scientific concepts and methods for this purpose and adapting them for their own purposes. And they do so. This is where methods from mathematics come into play. Especially at *ZEIT ONLINE*, the teams for investigative and data journalism as well as data visualization and interactive content are additionally supported by two internal *data scientists*, each having a mathematical background, who accompany the editorial work. This setup is unique in Germany.

In the following, we will give some practical examples, where data scientists at *ZEIT ONLINE* paved the way for the journalists behind the scenes.

Some examples from practice

Clustering, data analysis: Street names

In 2018, *ZEIT ONLINE* published a visual essay about street names. The main data source was the set of all major German streets in openstreetmap.org. The whole project, which won many prizes, would not have been possible without mathematical tools. Data was stored in a postgresql database and analyzed afterwards. A first step was the extraction of the core components of the street names by means of a heuristic (which assigned for example "Friedrich-Schiller-Platz", "Schillerstraße", "Schillerplatz", … to schiller). Then we focused on the location and form of the streets. Each street is stored as a set of sometimes only a few, sometimes hundreds of lines ("multilines"), sometimes even separate for each lane. The barycenters of the endpoints of these lines were used to compute a "center location" of each street, to further cluster all streets of the same core component. Moreover, the barycenter of the street-barycenters for each cluster has been compared with the barycenter of "Hauptstraße" ("Main Street"). This was done for each core component with a sufficient number of streets. "Hauptstraße" was chosen as a standard since almost every town has at least one street with that name. The aim was twofold: we wanted to know algorithmically if there are multiple clusters[a] of, say, schiller streets in Germany and if these streets have a similar distribution as "Hauptstraße" or not.

If the streets of one core component could be separated into more than one cluster and/or if their total barycenter differed significantly from the "Hauptstraße" barycenter, the corresponding core component was classified as "interesting" and shown to the editors. Thus, surprising clusters and patterns popped up automatically. An example was the Riesling-streets (see Figure 3). The editors also had to discard previously made hypotheses (for example a clear east-west-distribution for the rathenau-streets) — a nice example of hypothesis testing in data journalism. Data science also tested and discussed several approaches to rank streets by "curvyness". The idea

Figure 3: Three examples of unexpected patterns in street names. Shown are the barycenters (positions) of streets in Germany, different colors show different clusters. Limes streets (left) follow the course of the frontier of the Roman Empire (and had been separated in three clusters by the algorithm), riesling streets (right) are mainly found in one single cluster in the areas where Riesling grapes are grown, and welfen streets (middle) can mainly be found in three clusters in the former dominion of the Guelphs [14].

---

[a] Technically, we clustered with kMeans for a series of k between 1 and 10 until the sum of squared errors dropped under a threshold.

was to find the German street with the most curves, but it turned out that the definition of "curvyness" is difficult and that due to several reasons the data did not allow computation of any valuable curve-score, so this idea had to be dropped.

The project resulted in a loose collaboration and two small workshops with a team of social scientists and historians who focus on street names in Germany and Poland.

Visualization: Federal elections in Bavaria, opinion distances

The initiative abgeordnetenwatch.de is constantly surveying deputies of federal parliaments and the *Bundestag*. Before federal and state elections, all candidates get a questionnaire with yes/no/abstention questions, specifically for each election. For the federal elections in Bavaria in 2018, there were in total 20 questions. From the answers, one can derive distance relations between candidates (and, later, deputies) and get a complete distance graph with the nodes representing deputies.

After the election, the nodes of this graph were arranged by means of a spring embedding algorithm (see Figure 4). Additionally, we computed for this article how good the party CSU agrees with other parties, to estimate possible coalitions. This is an example of "invisible maths": Explaining mathematical methods, algorithms and possible pitfalls in data would have been straying far away from the main issue of this article, and in this case, the subject was small and of very short interest; a second technical explainer article would therefore probably not have found enough readers. However, in the background, the journalists indeed discussed several measures of distance and tested the visualization for robustness, since the algorithm is initiated randomly and produces slightly different visualizations with each run.

Database: Flights during Corona

In 2020, the flight tracking service flightradar24.com provided *ZEIT ONLINE* with data about 4.3 million flights worldwide. The data set was pushed into a PostgreSQL database and analyzed afterwards. Amongst other things, we checked which airplane types were registered at which times. This showed that after the decline in COVID-19 cases in March/ April, smaller planes (and in particular: not scheduled flights) were

So nah sind sich die bayerischen Abgeordneten

Je ähnlicher die jetzt gewählten Abgeordneten beim "Kandidatencheck" geantwortet haben, desto näher stehen sie in dieser Grafik zusammen.

Teilnahmequote: 70 Prozent

Quelle: abgeordnetenwatch.de / eigene Berechnungen · Daten

Figure 4:   Under the headline "That's how close the Bayarian deputies are to each other", *ZEIT ONLINE* presented a map of the new elected deputies in the Bavarian federal parliament after the election in 2018. The data basis was a questionnaire of abgeordnetenwatch.de before the election. The more similar the answers of the deputies had been, the nearer they where afterwards positioned on the map. The nodes are arranged by political distance, based on the 20 questions of abgeordnetenwatch.de [20].

responsible for the rising number of flights in summer 2020 [15]. The data was analyzed not only by plane size, carrier, and country, but also by distance of the flight, which had to be computed.

Although the mathematics behind the analysis was limited to pretty elementary computations of spherical distances on the surface of the earth, the database technology behind the analysis would not have been possible without mathematics, and only with a database the queries could be done in a realistic amount of time. An analysis of this size would certainly not have been possible with some table calculation software.

Matching: Europe talks

Often, publishers are much more than just news providers, they can also play a role of a host for communities. A good example is the projects "Germany talks" or "Europe talks". The idea is to connect people in

Germany or even all over Europe pairwise such that each pair has maximally different positions in a set of political questions, to discuss these questions and learn from each other.

This has been realized on many occasions now, for instance on May 11, 2019. About 16,000 Europeans met then at "Europe talks", both virtually and in person, to talk about politics. The project was an initiative of *ZEIT ONLINE* and was supported by 16 European media partners [18]. It was not the first approach of this kind: Already in 2017, *ZEIT ONLINE* organized "Germany talks" and in 2018, "Germany talks" took place in cooperation with several media partners. Without mathematics, none of these missions would have been possible, and each mission was a tremendous success.

The starting point is to have people answer a questionnaire of k yes/no questions to define a pairwise Hamming distance between them with values between 0 and k. Afterwards, people are paired under certain side conditions such that the net answer distance is as large as possible. In other words, we aim to compute a matching of maximal total weight.

In this size, however, this cannot be done optimally with a computing time of some minutes as is desired. So we implemented and used for this task an $(2/3 + \varepsilon)$-optimal approximation scheme for matchings from 2002 [16].

Integer linear programming: Die 49

A typical dichotomy in journalism is that often very general facts are wrapped into stories about individuals. An example would be an article about problems in the educational system of a country, motivated and told along the story of a single individual teacher. (However, each individual brings in their very specific problems, so this approach has its pitfalls.) Clearly, it would be nice to have a medium-sized set of individuals which is in certain aspects similar to the general population. This is the idea behind "Die 49".

More than 30,000 users of *ZEIT ONLINE* applied to become part of this group of 49 persons. Out of these, the data science department at *ZEIT ONLINE* computed a selection of 49 persons which are in 20 features as similar to the population of Germany as possible. The features were, amongst others: gender, size and location of the home town, household income, number of persons in the household, and sexual preferences.

This optimization problem was solved by means of an integer linear programming model. The final model had several million equations and inequalities and could be solved to optimality within 5 minutes on a PC notebook. It turned out that all 20 features could be satisfied *exactly*, meaning that the set of 49 persons contains for example 40 persons living in Western Germany, two persons from Berlin and seven persons from the Eastern part of Germany, and, at the same time, seven persons living in a small village and 16 living in a large town. Of course, cross dependencies could not be modeled (because of statistical reasons, and also simply because the underlying distributions are not known in general). So, for example, the group therefore does not contain the correct number of "gay men from large cities with low income". But "Die 49" can be seen as a first (and certainly interesting) approximation in that direction.

Conclusion

Mathematics as a subject is a niche in mass journalism. It certainly has entertainment value (for example in puzzles or as a "curiosity") and it fulfills an "advisor function" in the context of science or service journalism, that is, journalism which provides some kind of service to the readers to cope with a real live problem. More often, mathematical experts are used as universal experts who can "bless" and approve arguments, even on subjects outside maths, and give them more weight in discussions. The communication of content or ideas of current research, however, is almost non-existent in mass journalism, at least in Germany. One limiting reason is that recipients are only expected to have a knowledge level of high school mathematics at maximum, mainly not exceeding basic arithmetics. In fact, not many readers seem to be interested in a popularization of harder mathematical content. Even in the much larger English speaking market, math journalism like in *Quanta Magazine* or *plus.maths.org* live in a niche and can not survive without funding from foundations and public money.

Behind the scenes however, a different situation is developing. In newsrooms, a growing and increasingly professional number of data journalists and scientists are establishing mathematical methods and tools for editorial work, especially for the examination of large amounts of data. This development is only in the beginning phase and it can even lead to a better exchange with the scientific community in a new context.

References

[1] Mendick, H. (2015). *Mathematical Images and Identities: Education, Entertainment, Social Justice: Full Research Report ESRC End of Award Report.* RES-000-23-1454. Swindon: ESRC.

[2] Matzat, M. (2018). "10 Jahre Datenjournalismus: Gemischte Gefühle". *Datenjournalist.* https://www.datenjournalist.de/10-jahre-daten journalismus-gemischte-gefuehle/.

[3] Kepplinger, H. M. (2011). *Journalismus als Beruf.* VS Verlag für Sozialwissenschaften.

[4] Schwiesau, D., and Ohler, J. (2016). *Nachrichten — klassisch und multimedial — Ein Handbuch für Ausbildung und Praxis.* VS Verlag für Sozialwissenschaften.

[5] Fritsch, O. (2019). "Der Libero ist zurück". *ZEIT ONLINE.* https://www.zeit. de/sport/2019-05/fussball-bundesliga-libero-spielsystem-taktik-makoto-hasebe.

[6] Amrhein, V., Greenland, S., and McShane, B. (2019). "Scientists rise up against statistical significance". *Nature*, 567, 305–307.

[7] Meyer, P. (1991). *The New Precision Journalism.* https://carolinadatadesk. github.io/pmeyer/book/.

[8] Gray, J., Bounegru, L., Chambers, L., *et al.* (2012). *The Data Handbook 1.* European Journalism Centre, https://datajournalism.com/read/handbook/ one.

[9] Weinacht, S. and Spiller, R. (2014). "Datenjournalismus in Deutschland — Eine explorative Untersuchung zu Rollenbildern von Datenjournalisten". *Publizistik*, 59, 411–433.

[10] Breda, A. (2019). *Objektive Diagramme gegen die Vertrauens- und Finanzierungskrise? — Inhaltsanalyse datenjournalistischer Visualisierungen.* Master Thesis in Journalism, University of Leipzig.

[11] Silver, N., *et al.* (2020). *How (Un)popular is Donald Trump?* https://projects. fivethirtyeight.com/trump-approval-ratings/.

[12] *The Economist.* (2021). "There have been 7m-13m excess deaths worldwide during the pandemic". https://www.economist.com/briefing/2021/05/15/ there-have-been-7m-13m-excess-deaths-worldwide-during-the-pandemic.

[13] Weinacht, S. and Spiller, R. (2013). "Wie wissenschaftlich ist Datenjournalismus?". *WPK Quaterly*, 1, 14–15.

[14] Venohr, S., *et al.* (2018). "Straßenbilder Mozart, Marx und ein Diktator". *ZEIT ONLINE*. https://www.zeit.de/feature/strassenverzeichnis-strassennamen-herkunft-deutschland-infografik.

[15] Venohr, S., *et al.* (2020). "Wir haben die Parkposition erreicht". *ZEIT ONLINE*. https://www.zeit.de/mobilitaet/2020-08/flugverkehr-flightradar24-flugbewegungen-coronavirus-anstieg-flugzahlen.

[16] Drake, D. E., and Hougardy, S. "A simple approximation algorithm for the weighted matching problem". *Information Processing Letters*, 85(4), 211–213.

[17] o.A., *Die Statistische Methodik von Civey*. civey.com/whitepaper/, o.J.

[18] Faigle, P., Loos, A., *et al.* (2019). "Europa, im Streit vereint". *ZEIT ONLINE*. https://www.zeit.de/gesellschaft/2019-05/europa-spricht-gespraech-meinungen-teilnehmer-analyse.

[19] Galtung, J., and Ruge, M. H. (1965). "The structure of foreign news". *Journal of Peace Research*, 2(1), 64–91.

[20] Loos, A., Tröger, J., and Venohr, S. (2018). "Mit wem kann die CSU am besten?". *ZEIT ONLINE*. https://www.zeit.de/politik/deutschland/2018-10/regierungsbildung-bayern-koalitionspartner-csu-gruene-freie-waehler.

https://doi.org/10.1142/9789811253072_0019

Chapter 18

From Education to Edutainment in Mathematics

Thomas Vogt

Math Media Office of the
German Mathematical Society,
Freie Universität Berlin, Germany
vogt@mathematik.de

From education to edutainment in mathematics — a survey on science, technology, engineering, and mathematics (STEM) activities beyond the public school system in Germany and trends in science communication with an emphasis on math in Germany by Thomas Vogt.

A lively scene of STEM activities and initiatives has arisen in Germany beyond the public school system during the last two decades. In this article, we make an inquiry on existing activities and initiatives in STEM subjects in Germany with a focus on the youth, Germany's "Science Years" ("Wissenschaftsjahre"), the "Year of Mathematics" (2008), and its aftermath. We try to give an exhaustive outline on the existing activities but will not evaluate what's being offered. We aim to show if or how the offer completes public education and ask for reasons for this extensive offer beyond the public school system that reaches from education to edutainment. Additionally, we outline trends in science communication in Germany with an emphasis on mathematics during the past 20 years. That's why some of the sources below date back to the year 2000 and before. Links to online material for the various activities and initiatives can be found in the footnote section, the most recent entry being from March 2022. Printed sources may be found in the reference section. Some sources appear in both sections when the access via internet is free of charge.

PUSH and PISA

The British memorandum on the Public Understanding of Science and Humanities (PUSH) in the 1990s had a strong impact in Germany. This can be seen, for example, in the "Symposium Public Understanding of Science and Humanities — International and German Perspectives", that took place in 1999 and which led to the foundation of Wissenschaft im Dialog (WiD), which means "Science in Dialogue", in the year 2000.[a]

---

[a] https://www.wissenschaft-im-dialog.de/en/about-us/.

With this initiative, national research associations like Helmholtz, Max Planck, Fraunhofer, Leibniz Society, and the science foundation Stifterverband joined forces to promote the natural sciences (and humanities) beginning with physics in the year 2000. WiD developed several new communication formats over the following years, see examples below.

Peters, Lehmkuhl, and Fähnrich give a good overview on the long tradition of communicating science in Germany, published by the Australian National University ANU [1].[b]

A crucial driving force for the strong impact of PUSH, reforms in education, and many activities in science communication in Germany described below was the so-called PISA shock in Germany: In 2000, OECD's Program for International Student Assessment (PISA) had revealed competences below OECD average in reading, mathematical literacy, and other STEM subjects throughout Germany [2].[c] After the Third International Mathematics and Science Study II (TIMSS II, 1997[d]), this was the second international study that disappointed German people and politics, which had felt a certain pride for the German educational system up until then.[e]

These poor results in the TIMSS and PISA studies and the lack of students enrolled in STEM subjects at universities were seen as unacceptable for Germany as a location known for (mostly technical and chemical) industry, and it led to an intensive public debate on reasons and solutions for the problem.[f]

The (TIMMS II and) PISA shock caused ripples in various areas in Germany, in state education, and throughout society as a whole. Almost all federal states enacted reforms for elementary and secondary schools and education budgets were about doubled within 10 years. All measurements

---

[b] https://press-files.anu.edu.au/downloads/press/n6484/html/ch14.xhtml.

[c] https://www.oecd.org/education/school/2960581.pdf.

[d] https://timssandpirls.bc.edu/timss1995i/timsspdf/tr2chap1.pdf.

[e] https://www.oecd.org/about/impact/.

[f] https://www.kas.de/c/document_library/get_file?uuid=15147722-43a0-63e9-993e-0c44c26da633&groupId=252038.

taken together lead to a certain success within 15 years.[g] For reasons of brevity, we will not say more about this in this article. Instead, we would like to focus here on the non-school sector.

Research organizations, ministries of Education, and the business community were all considering how to make STEM subjects more popular and successful in Germany. The situation was precarious in several respects: there was already a shortage of teachers for STEM subjects, companies had (and still have) increasing problems finding trainees who have sufficient knowledge of mathematics for technical professions, and large companies fear a lack of highly qualified junior staff for the export industry.

This circumstance called a number of new players with different interests onto the scene. Almost all initiatives had in common the desire to improve the mathematics knowledge of students in particular. Research institutes of the Helmholtz Association, for example, but also universities and even large clinics set up so-called School Labs ("Schülerlabore") as a new form of extracurricular places of learning where school material could be practiced in the course of practical application. Universities invited students to so-called Student Universities ("Schüleruniversitäten") in their hallowed halls and engaged in the promotion of the gifted, such as with student competitions or student circles. Some came up with completely new formats, such as the Technical University of Berlin (TU Berlin) with the Math Advent Calendar of the DFG Research Center MATHEON. For this calendar, MATHEON doctoral students write original math problems that relate to their research topics, and they can be opened and solved from December 1 to December 24 via the daily "little doors" of a digital Advent calendar (level 3). In an associated forum, relevant mathematical problems can be discussed. At the end, high-quality prizes await the winners. The tasks of the MATHEON calendar (today MATH+) go well beyond school mathematics and are aimed at mathematically gifted high school students and first-year students of mathematics.[h] Participation numbers increased in this level over the years to around 10,000 people, half of whom are adults interested in mathematics. In today's three levels for classes 4–6 and 7–9

---

[g] https://www.oecd.org/about/impact/germany-pisa-shock.htm.

[h] https://www.mathekalender.de/wp/.

("Mathe im Advent") and 10+ (MATH+) of the Math Advent Calendar, there were 195,000 participants in December 2021 (sum of players in all levels[i]).

In the following, we would like to briefly explain the new extracurricular learning venues, the School Labs.

School Labs

The School Labs are sometimes more hands-on exhibitions and sometimes more makerspaces but always give the visitors the opportunity to practice and experiment.

The idea behind the School Labs was to complement the theoretical background learned at school with a hands-on experience. The belief was that college students would better understand what teachers had talked about in class and what the practical implementations of knowledge in STEM subjects could be. With an excursion to a place outside school or college, the hope was that learning routines would be broken, attention would be risen and a "Eureka!" moment would occur and thus help students to become interested in the subject and to better understand what certain STEM subjects were "really" about. This simple scheme worked relatively well, judging by the numbers of visits to the School Labs. Many Labs were booked out by classes several weeks in advance. The offer was mostly free of charge for school classes.

The operating institutions also profited. Research centers with different thematical focuses had a new opportunity to welcome pupils and college students at their facilities. Examples are the institutes of Helmholtz, Fraunhofer, the Leibniz Association, and Max Planck Society as well as large universities and even hospitals.[j] The School Labs gave the research institutes the opportunity to present themselves to the public, especially to the youth and to interest them in their research topics, mostly STEM subjects. Schools also profited by upgrading their classes with an excursion to a research facility, sometimes even creating curiosity in pupils with a previously low interest in certain STEM subjects.

---

[i] https://www.mathe-im-advent.de/en/.
[j] https://www.lernortlabor.de/ueber-schuelerlabore.

This win-win situation led to the School Labs being a huge success. The School Labs of Helmholtz Society[k] alone had around 100,000 visitors in 2019, said the "Gläsernes Labor" on its website.[l] This was also possible because the topics in many School Labs would refer to the curriculum at school and thus made it easier for teachers to integrate them into the classroom. For example: The PhysLab at Freie Universität Berlin (FU Berlin) developed a learning cycle of experiments on "swimming, floating, sinking" (topic upswing) in 2006 and they recorded that it was attended by around 500 children, 800 teachers (teacher training!), and 300 students (mostly in didactics). The contents of the sessions correlate with the curricula of Berlin's elementary schools.[m]

Examples for activities at some School Labs are:

At the Math Lab of TU Berlin, the traveling salesman problem was explained via a wooden map of Germany with holes drilled at the locations of major cities in which little sticks could be inserted. The students could now fiddle with a string along the sticks and find the shortest way to pass by 10 of the large cities via trial-and-error (Figure 1). In the Life Science

Figure 1: Traveling salesman problem: finding the shortest way to pass by 10 large cities in Germany via trial-and-error. Math exhibition on "MS Mathematik". Copyright Ilja C. Hendel/Wissenschaft im Dialog.

---

[k] https://www.helmholtz.de/en/transfer/school-labs/.

[l] https://www.glaesernes-labor.de/de/news/helmholtz-bilanz-2019.

[m] http://www.physik.fu-berlin.de/physlab/.

Learning Lab ("Gläsernes Labor") at the clinical campus of Berlin Buch, students can isolate DNA out of the cells of their oral mucosa. This School Lab also offers classes in biology for final exam students of biology and teachers training in genomics.[n] The School Labs reflect the desire to teach school subjects outside school in a practical way while constituting another (practical) way of learning.

Stakeholders

By implementing the School Labs, new protagonists and sponsors emerged on the scene, including hospitals, foundations and companies. Another important sponsor became the Federal Ministry of Education. This is interesting from a German perspective, because educational issues, such as the curricula of schools, colleges, and universities, are in the hands of the German federal states ("Länder") and thus are financed regionally by the respective state ("Bildungsföderalismus"). The Federal Ministry of Education has only a consultative and moderating role concerning public schools and colleges as one of 17 members of the Standing Conference of the Ministers of Education and Cultural Affairs ("Kultusministerkonferenz", KMK[o]), i.e., the 16 federal states plus the Federal Ministry of Education. KMK gives recommendations, not directives. On the other hand, the national government finances Germany's research activities via the budgets of the German Research Association (Deutsche Forschungsgemeinschaft DFG), the Helmholtz Association and partly Leibniz and Max Planck institutes. So WiD may also be understood as a construction for the national government to finance science communication (as a higher-ranking activity beyond the responsibility of the federal states).

This distributed responsibility of formal education issues in Germany between the federal states and the national government gives a lot of room for educational and science communication activities on a regional level, such as science festivals, science museums, science centers, and Houses of Science, see paragraphs below.

---

[n] https://www.glaesernes-labor.de/de/learning_lab?domp=glab.

[o] https://www.kmk.org/kmk/information-in-english.html.

The activities outlined above (school labs, student circles, student competitions) are predominantly science education measures supplementing school lessons. The primary goal of the scientists involved was to provide students with "first-hand" knowledge about their subject. This goal is also seen in most of the statements made and projects outlined in the handbook *Raising Public Awareness of Mathematics* on a European level [3]. In terms of target groups, the group of people who were addressed were already interested in mathematics, often even considered mathematically gifted. This has its justifications. However, from the point of view of the organizers of the Science Year of Mathematics, the problem was a different one, namely the lack of interest in mathematics in many people, sometimes fear of mathematics, often from a young age on. In any case, the moderate knowledge of students in mathematics, determined in the PISA studies and other studies from the educational and social science fields, were interpreted in this way [4].

Science Years

Compared to the activities of universities and research institutes described above, the Science Years represented a real turning point: For the first time, the focus of a major science communication project was not on imparting knowledge, but on having fun. The declared aim was to interest, and indeed inspire, broad sections of the population in the STEM subjects. After all, many scientists, especially mathematicians, perceived prejudices against their subject and person, which some studies on the image of mathematicians confirmed. Susan H. Picker and John S. Barry, for instance, let pupils from half a dozen countries aged 12–13 years draw their image of a mathematician showing that "(...) with small cultural differences certain stereotypical images of mathematicians are common to pupils in all of these countries and these images indicate that for pupils of this age mathematicians and the work that they do are, (...) invisible." and worse: Some pictures even show mathematicians with a shotgun [4].

For mathematics, this meant that even less mathematically-savvy groups should gain interest in mathematics. This should be achieved not from a defensive stance, but by positively highlighting the special importance of mathematics for society. Mathematicians and an advertising agency came up together with numerous examples in which mathematics plays a central role in society, e.g.:

- The importance of mathematics in many other subject areas, even beyond STEM, such as economics and the social sciences; a fact that is still being studied and communicated today, see this Delphi study of IPN 2021.[p]

- In addition, there were many practical examples of the importance of mathematics in everyday life in medicine (e.g., image-forming procedures), the stock exchange (e.g., stock trading), chemistry (e.g., vaccine and drug production), as well as for large industrial companies such as aviation and car development, the internet (e.g., online banking and internet security), data compression (e.g., data transfer, Big Data, music files), weather forecasting, radio cell location, and much more.

- The occasional reference was also made to intuitive "math skills" in the broadest sense, such as geometric skills for parking a car, or of tying a tie like Fink and Mao showed [5].

Several formats were used to try to challenge people's play instincts, for example, with Sudokus and other puzzles and quizzes. The goal was to get people to actively participate and invite them to mathematical activity. This intention is still embodied today in the IMAGINARY project in a sophisticated form, for a full overview of this project see Chapter 8 of this handbook. An example of this is seen in IMAGINARY's license-free software SURFER, which makes it possible to create a visual effect by playfully changing a formula. "Just invent an equation and play with it! Enjoy doing and learning maths visually!" it says on their website.[q] Even though the word "learn" still appears in this motto, the approach is primarily a playful one, namely to create visually attractive algebraic surfaces. At the same time, it creates a certain sense of wonder, and in the best case, even interest in mathematics (i.e., algebraic surfaces!), is generated. Thousands of students experimented with SURFER during the

---

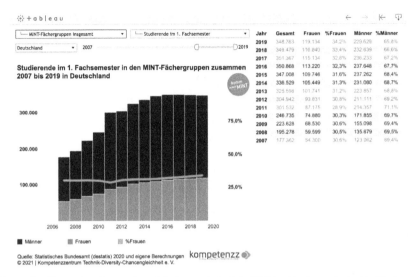

Figure 2: Shows that the number of first year students in STEM subjects nearly doubled between the years 2007 and 2019 (column "Gesamt", i.e., total). The percentage of female students ("Frauen") rises continuously since the year 2006. Numbers provided by Statistisches Bundesamt (destatis.de) 2020 und Kompetenzzentrum Technik-Diversity-Chancengleichheit e. V., 2021. https://www.komm-mach-mint.de/service/mint-datentool.

Year of Mathematics and created thousands of imaginative shapes, which were submitted to online competitions and awarded prizes, for example as part of media partnerships with the science magazine *Spektrum der Wissenschaft* and *ZEIT ONLINE*.

But now back to the Science Years and their communication formats. They allowed to perfectly address broad target groups and to achieve the communication goals mentioned above, as evidenced by the high number of participants, very good media coverage, and even increasing numbers of first-year students in STEM subjects in the early 2000s (Figure 2).[r]

From 2000 onwards, the Federal Ministry of Education and Research (BMBF) and the Stifterverband jointly proclaimed a Year of Science for a specific subject each year, starting with the Year of Physics (2000). While WiD was primarily concerned with the development of new communication concepts and their realizations, the expert associations contributed the content, i.e., the institutes of Helmholtz and the Leibniz

---

[r] https://www.komm-mach-mint.de/service/mint-datentool.

Association, of Fraunhofer and the Max Planck Society, thus playing a leading role. The largest and most important communication formats of the Science Years were Science Summer ("Wissenschaftssommer") and the exhibition ship MS Science ("MS Wissenschaft"). In 2008, it was mathematics' turn for a Year of Science. More on this below.

Science Summer (2000–2012)

Cities had to apply to host Science Summer and mostly large, regional cities won the bid, such as the state capitals of Mainz (2003 and 2011), Stuttgart (2004), Munich (2006) and Essen (2007). Science Summer was a science festival where exhibits and science shows were offered in pavilions and public spaces (in large central squares, pedestrian zones, etc.), mostly free of charge. "Hands-on science" was the motto. Consequently, the staff was recruited mainly from the research community and, over the years, increasingly from the entertainment sector as well. Doctoral students and postdocs, for example, presented their research in a "science slam", but science-related cabaret artists and bands also provided entertainment. Science Summer became a destination for the whole family.[s]

The format of Science Summer was so successful that it was the inspiration for the invention of other science festivals, for example, the Long Night of Science ("Lange Nacht der Wissenschaften"). The Long Night of Science, which runs from early evening until at least midnight, adopted the idea of a summertime mix of presenting research and entertainment in the form of a city- or neighborhood-wide science festival for broad segments of the population. Supplemented by various outdoor activities of a musical and culinary nature, the Long Night of Science became an experience for young and old. The fact that this science festival ran independently from the Year of Science and had its theme meant that the Long Night of Science could take place every year according to local wishes and circumstances. For numerous science institutes, the Long Night of Science became a fixed part of the annual calendar of events for science

---

[s] https://www.wissenschaft-im-dialog.de/projekte/abgeschlossene-projekte/wissens chaftssommer/.

and culture, for example in the Berlin–Potsdam region.[t] Since many large cities in Germany have a Long Night of Science today, WiD stopped its Science Summer project so as to not overshadow these regional activities (having successfully played an initializing role).

MS Science

In addition to Science Summer, the ship exhibition MS Science (new from 2002 on) was an important communication format of WiD. It had an exhibition thematically related to the respective Science Year onboard and thus could come as a floating exhibition across Germany's inland waters to people all over the country and in some years even as far as Austria. The onboard exhibition mainly had interactive exhibits (Figures 1, 3–5). The exhibits were each contributed by WiD's sponsoring institutions.

Figure 3:   School class coming from math exhibition on math ship "MS Mathematik". Copyright Ilja C. Hendel/Wissenschaft im Dialog.

---

[t] www.langenachtderwissenschaften.de.

Figure 4: Brachistochrone experiment in math exhibition on math ship "MS Mathematik". Copyright Ilja C. Hendel/Wissenschaft im Dialog.

Figure 5: Scrabble experiment in math exhibition on math ship "MS Mathematik". Copyright Ilja C. Hendel/Wissenschaft im Dialog.

The exhibition was therefore also a showcase of the relevant facilities and institutes, which increased the internal acceptance of the exhibition as well. The ship stopped in larger cities along the rivers. Its stay of 1–2 weeks was advertised locally and made it possible for entire school classes to visit the ship (Figure 3). In this respect, as with Science Summer, the idea was to reach out to people and get them interested in science through unusual formats. The number of visitors to the MS Science was enormous by German standards. For example, the exhibition had 117,000 visitors in the Year of Earth Sciences (2002), 118,000 visited the MS Mathematik, and 128,000 visited the MS Seas and Oceans (2017) between May and October (numbers provided by WiD).

Year of Mathematics

We will now elaborate on the Year of Mathematics for several reasons; firstly, because it corresponds to the focus of this article, secondly, because it represented a highlight in the long series of Science Years and lastly because it had a lasting impact on the press and public relations work of the German Mathematical Society ("DMV"). The success of the Year of Mathematics is evidenced by the high number of participants in its events, as well as the great (predominantly positive) media response, with around 100 articles and other pieces each month about mathematics and the Year of Mathematics in regional and national media.

There is some information on the success of the Year of Mathematics on the internet.[u] Since it is all in German, we summarize here the most important dates and figures on the Year of Mathematics:

500 partners from science, culture, art, and industry organized together 762 events all over Germany. They took place in about 140 towns. 150 of the events were part of the network project "Mathematik vernetzen" between universities and local schools with the DMV, the Telekom foundation, and the teachers' society MNU as genuine partners.

---

[u] https://www.wissenschaftsjahr.de/2008/.

About 400,000 pupils were reached via information material and the so-called math suitcases ("Mathe–Koffer"), a bundle containing math games and instruction material for school children. Around one million pupils took part in 30 math competitions reaching from existing national and international math competitions like IMO to new online competitions like ZAL (25,000 participants), the Math calendar (50,000), and Math Kangaroo (750,000 pupils). A special competition asked for ideas to teach math in a new, supposedly better way, named "Experience math!" ("Mathe erleben!"). 205 projects were awarded €5,000 each for realization. The 20 best received a €10,000 prize, for instance the map exercise "Planspiel Stadt", which was realized in 23 cities together with the national city association ("Deutscher Städtetag"). A math film festival took part in 100 cities in Germany.

Exhibitions of IMAGINARY, "12 sind Kult" (12 is cult), "Zahlen bitte" (numbers please), and the floating exhibition MS Mathematik (118,000) had all in all 400,000 visitors, the "Wissenschaftssommer" in Leipzig an additional 100,000 participants, including Leipzig's first Long Night of Science on June 28, 2008.[v]

The author can only speculate as to why the Year of Mathematics was so successful. Therefore, only three potential reasons will be mentioned here: one is that mathematics is a (core) school subject and thus every child, young person, and their parents have an opinion about the subject. Regardless of whether this opinion is positive or negative, the subject polarizes and invokes strong emotions, which is an advantage for successful communication. Everyone can or wants to have a say, and often does. Then the seemingly boundless enthusiasm of mathematicians from all over the country was extremely helpful. In many places, mathematicians and non-mathematicians developed ideas together and realized them with their own resources. One example from the city of Bremen: A bakery, in cooperation with the local mathematical institute, had paper bags printed with the basics of knot theory and sold their pretzels in the bags. Actions like these (financed locally) were then reported on by the regional media.

Another reason for the success of the Year of Mathematics could have been the structural or organizational nature of the Year of Science 2008. Mathematics spoke with one voice because relevant associations, such as the teachers' association "MNU" and the Society for Didactics of Mathematics ("GDM"), let the DMV speak for them. So, for mathematics and for the community, there was only one contact, namely the DMV, which was represented by its young, charismatic president Prof. Günter M. Ziegler, who in many respects contradicted the cliché of a mathematician. Eloquent and quick-witted, Ziegler soon became a darling of the media, making it into popular TV quiz shows.

Another strong partner in the Math Year was the Deutsche Telekom Foundation, which under its chairman former Foreign Minister Dr. Klaus Kinkel made mathematics its funding priority because it plays a key role in studying and training for STEM professions that are of particular importance to its parent company Telekom. Kinkel strongly promoted the Year of Mathematics in his environment and acted as a promoter to get other foundations, ministries, and companies involved. The Telekom Foundation itself also contributed substantial funds to further projects in the Year of Mathematics over the course of the year.

WiD and BMBF were set as partners and were also an asset with their many years of experience in planning the Science Years (Figure 6).

Figure 6: The partners of the Year of Mathematics 2008: Prof. Günter M. Ziegler, president of DMV, Prof. Gerold Wefer, president of WiD, Federal Minister of Education Dr. Annette Schavan and Dr. Klaus Kinkel, president of Deutsche Telekom Stiftung (left to right). Photo by Amin Akhtar, 2008.

Additionally, the commissioning of one of the most successful advertising agencies in the country at the time, Scholz & Friends, was another advantage. The agency had lots of experience in campaigning and event management, too.

Advertising Agency

Right at the beginning of the Year of Mathematics, the agency Scholz & Friends was commissioned to develop and, if necessary, implement various activities. These included the poster campaign, the ambassador and Math Maker campaigns, and hosting the internet portals, all discussed below. For the creation of the contents, the editorial office in the Year of Mathematics was founded and supported by the "Math Contents Back Office" (MCBO), a branch office of the Institute of Mathematics that Prof. Ziegler has chaired at TU Berlin, now at FU Berlin. The MCBO, which still exists today as the Mathematics Media Office ("Medienbüro Mathematik"[w]) and is managed by the author, had the task of creating topic-specific content with feedback from the mathematical community and, together with the agency, feeding it into the communication process in a media-effective manner.

The happy coincidence of a large number of the above-mentioned players and the great enthusiasm of the mathematics community led to a fireworks display of activities. Of the 762 events and projects that took place throughout Germany in 2008, only a few particularly successful and instructive ones are mentioned here as best practice examples.

Websites and Topics for Different Target Groups

Right from the start, two internet portals were set up for two different target groups: The internet portal *jahr-der-mathematik.de*[x] was aimed at a broad public and provided bundled information about all events and activities in the Year of Mathematics. It was aimed at adults of all ages, teachers, students, and representatives of the media. The program was correspondingly colorful: On the homepage, a puzzle of the week was offered along with daily news from mathematics and, in the form of topic

---

[w] https://www.mathematik.de/kontakt-dmv.

[x] https://www.wissenschaftsjahr.de/2008/.

dossiers for the media, in-depth topics from mathematics that dealt with the omnipresence of mathematics in today's world, for example "Mathematics and Medicine", "Mathematics and Sport", and "Mathematics and Space Research". On the occasion of the 13th European Soccer Championship in 2008, another dossier was offered on the topic of "Mathematics and Soccer". It ranged from simple methods for calculating the penalty area and the penalty spot to sophisticated calculations of the chances of victory for individual groups and countries using combinatorics. The dossiers were developed by the editorial office in the Year of Mathematics and the MCBO, founded by Prof. Ziegler, see above.

The second internet portal called *du-kannst-mathe.de* referred to the slogan "You can do more math than you think!" The pages were aimed specifically at young people and trainees with topics related to mathematics in everyday life, school and training. For example, on intuitively "finding" the right angle when striking at a soccer/handball goal, skateboarding, tying ties (topology), listening to music (possible due to digital compression methods), or parking a car. Posters on these topics were produced and put up nationwide.

Ambassadors for Mathematics

Another valuable idea to come from the agency was to recruit celebrities who have or had a connection to mathematics and would speak out as ambassadors in favor of mathematics. This included soccer coach Mirko Slomka, who studied mathematics and physical education to become a teacher before becoming a soccer player, then a coach. He said that mathematics plays a decisive role for him in almost all areas of life.

Gender and Diversity

In the Year of Mathematics, a targeted search was made for women who had studied mathematics and could serve as role models for girls and young women. The aim was to counteract the widespread prejudice in Germany at the time that mathematics was a male discipline.

An example of one prominent ambassador was Barbara Meier, winner of "Germany's Next Topmodel Award" in 2007. Even as an aspiring model, she said she wanted to pursue her mathematics studies. The mathematics student and musical actress Anna–Maria Schmidt, from the popular ZDF casting show "Musical Showstar 2008", said something alike publicly.

Activation Campaign

Another idea of the advertising agency that proved to be extremely successful was the selection of Math Makers of the week. The idea was to make voluntary engagement in mathematics visible and to literally give it a sympathetic face every week. In other words, to show not only celebrities but also "ordinary" people who do a great deal for mathematics in their everyday lives and especially in their spare time. For example, active or retired math teachers who also offer a chess club or train students for math competitions, mathematics students who offer free math tutoring in their neighborhood or people who make art or poetry inspired by mathematics in their free time. Interested people could apply to be a Math Maker ("Mathemacher") via the website by briefly describing their engagement activity and where selected by the editorial office. In this case a short, standardized interview was conducted and the completed statement "Mathematics counts because ..." was added to the online portal with a photo of the person. This activation campaign was so successful that about 900 Math Maker applications were collected during the year, more than could be considered. Because of the high demand, the Mathematics Media Office, the follow-up institution of the MCBO at TU Berlin, later at FU Berlin, decided to continue the Math Maker campaign after the Year of Mathematics at a lower frequency by awarding the title "Math Maker of the Month" in the name of Deutsche Mathematiker-Vereinigung, which still continues today.[y]

Activating the Play Instinct

Math puzzles and quizzes proved to be extremely popular in the Year of Mathematics. The weekly changing math puzzles on *jahr-der-mathematik. de* led to high click numbers. Puzzles and brainteasers were also in high

---

[y] https://www.mathematik.de/hochschule-beruf/mathemacher-innen.

demand from the media; for example BILD newspaper, Germany's largest-circulation print medium with eight million copies per day, printed a multi-page "Big Math Puzzle" in summer 2008.

In the case of the Math Quiz, Prof. Ziegler's version was particularly successful: He usually used the popular format of "Who Wants to Be a Millionaire?", in which a question is asked and four answers (A, B, C, D) are offered for selection, which sound more or less improbable. Therefore, the solution almost always creates a surprise effect. The Math Quiz by and with Prof. Ziegler was a highlight of almost every major event in the Year of Mathematics (Figure 7).

More on the Lessons Learned

For the DMV, the Year of Mathematics meant professionalizing its press and public relations work. On the one hand, the DMV learned a great deal from the above-mentioned agency and from testing various communication formats. Furthermore, the DMV was able, not least through the continuation of the MCBO, to further nourish the wave that

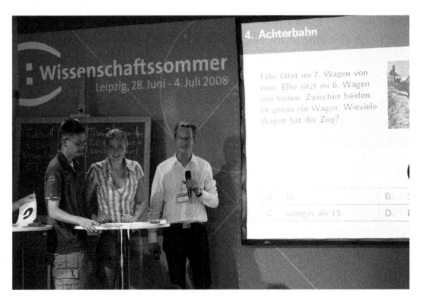

Figure 7: Günter Ziegler's popular math quiz in a "Who Wants to be a Millionaire?" style at the Science Summer in Leipzig 2008. Copyright Ilja C. Hendel/Wissenschaft im Dialog.

the math PR year had generated and to maintain interest in the subject with continued activities, as evidenced by a good media response to this day and increasing numbers of first-year students in STEM subjects (Figure 2).

In addition to the Math Maker award, the Abitur Award Mathematics ("Abiturpreis Mathematik") has also been awarded since 2008, with more than 3,000 prizes annually since 2009 until today.[z] The Abitur Award Mathematics may be awarded by any school in German-speaking countries and at German schools abroad leading to the degree Abitur on behalf of the DMV. It is awarded to high school graduates who have achieved a good or very good grade in the Abitur mathematics exam and who are also committed to mathematics, for example with tutoring and chess clubs or similar activities for students. The award consists of a certificate, a 1-year free membership in the DMV and a book prize donated by the publisher Springer for this purpose. To date, the book in question is *Pi & Co*, edited in the Year of Mathematics by the mathematics professors Behrends, Gritzmann and Ziegler [6]. The awarding of the Abitur award is overseen by the DMV Network Office School — University ("DMV-Netzwerkbüro Schule — Hochschule"), which aims to better connect schools to universities and ensure an easier transition from school to university in the long term. The Network Office is also a result of the Year of Mathematics and received substantial funding from the Deutsche Telekom Foundation in 2009 and thereafter. Together with the DMV Mathematics Media Office, it is still based in Prof. Ziegler's working group at FU Berlin.[aa]

Impact of the Science Years

Last but not least, the Science Years offered the opportunity to try out new formats of science communication. For example, for Physics in the Department Store ("Physik im Kaufhaus") in 2000, physicists presented physics experiments in the midst of a department store and tried to engage

---

[z] https://www.mathematik.de/dmv-blog/2921-schulen-zeichnen-3250-preistr%C3%A4ger-innen-mit-dem-abiturpreis-mathematik-2021-aus.

[aa] https://www.mathematik.de/kontakt-dmv.

in conversations with customers about research topics. Other formats in subsequent years were Science Station, in which scientists attempted to enter into a dialogue about science with travelers waiting for their trains in large train stations, and the science festivals, where scientists went to public spaces like pedestrian zones and marketplaces. The idea behind these formats was the same; science leaves its ivory tower and reaches out to people, even more so in places where citizens would not expect it, creating a certain element of surprise and the opportunity for casual discourse on science. The great popularity of these formats, especially the science festivals, shows that it worked well, see above.

Generally speaking, it can be said that the Science Years and the new communication formats developed by WiD and the associated professionalized science communicators have promoted interest in science in Germany over many years. This is shown by the numerous science festivals and activities that have emerged in parallel and in succession throughout the country. For example, the Science Market in Mainz, the Long Night of Science in several cities, the Houses of Science and numerous Days of Mathematics. These are sometimes more mathematics competitions for schoolchildren as in Berlin, Chemnitz, and Koblenz, sometimes more advanced training days as in the Dortmund Math Day, or even a kind of science market as in Freiburg or Mainz (which took place for the 18th time in 2019). Large science centers for science communication and citizen dialogue emerged also, see the presentation below. There was a boom in science TV programs and science broadcasts and podcasts on air, like Quarks and Co., W wie Wissen, Leonardo, Synapsen etc. and "Wissen macht Ah!" for children. For rising numbers of students in STEM subjects see Figure 2.

Obviously, it is not possible to prove that this increase in interest in science is an effect 100% dependent on the Science Years. What we can assert is that science topics are popular in Germany and that people generally have great confidence in science. This has been impressively demonstrated by the handling of the Fridays for Future demonstrations and the COVID-19 crisis in Germany, where politicians and serious media explicitly refer to the "research results of science". During the COVID-19 crisis, the Robert Koch Institute (RKI) and the Paul Ehrlich Institute (PEI) in Germany were and are considered extremely trustworthy. Conversely, the majority of the scientific community backed the concerns of the Fridays for Future movement by founding the supporting initiative Scientists for Future, and numerous professional societies also backed the youth's call for more climate protection, including the DMV.[ab,ac]

Of course, some science communication initiatives and venues were created before the year 2000, including Germany's first science center, the Spectrum, which opened in 1983 on the grounds of the Deutsches Technikmuseum Berlin, and was modeled after the Exploratorium in San Francisco, and the "Universum Bremen", opened in the year 2000.

One outstanding example of a center dedicated solely to mathematics will be described in more detail here, namely the Mathematikum in Giessen.[ad] The first — by its own admission — "participatory museum of mathematics" in Germany owes its existence to Albrecht Beutelspacher, professor of geometry and discrete mathematics at the University of Giessen (1998–2018). In addition to research and teaching, Beutelspacher has always been committed to the comprehensible presentation of mathematics and has written numerous popular science books on mathematics, such as "Mathematik für die Westentasche" (Mathematics for Your Pocket) or "Geschichte der Mathematik" (History of Mathematics). The first exhibits were built by students of mathematics, which were shown in an exhibition titled "Mathematik zum Anfassen" (Mathematics to Touch) at the Institute for the Didactics of Mathematics at the University of Giessen. In the course of the repeated exhibition, which was met with a good response, the idea of founding a mathematics museum was born. Beutelspacher founded a sponsoring association and successfully promoted his project in his town and state. In 2002, the Mathematikum was officially opened by the Federal President. Beutelspacher became the first director of the Mathematikum, further developed the permanent exhibitions and expanded the building. This soon made it possible to hold special exhibitions and events related

---

[ab] https://www.mathematik.de/presse/2618-naturwissenschaftliche-fachgesellschaften-zum-klimawandel-%E2%80%9Eh%C3%B6rt-auf-die-wissenschaft-%E2%80%9C.

[ac] https://bit.ly/3jyzM8k.

[ad] www.mathematikum.de/en/.

to mathematics. In 2009, the Mini-Mathematikum was created for children aged four to eight. The Mathematikum contains more than 170 mathematical experiments and exhibits to participate in, for example, a Hall of Mirrors, a Foucault pendulum, building Leonardo bridges, and 2D and 3D puzzles, on a good 1,200 m² of exhibition space. A particularly popular exhibit in the permanent exhibition is slipping into an oversized soap bubble (Figure 8). The references to mathematics are always explained in an understandable way. In addition, there are special exhibitions and talk formats such as "Beutelspacher's Sofa", in which the institute's director

Figure 8: Bubble soap experiment at the Mathematikum in Giessen. Copyright Mathematikum/ Rolf K. Wegst.

talks to invited guests, from fields such as science and politics, about mathematics and its impact on society.

The Mathematikum contains very educational exhibits, but also those that focus on the fun factor, such as the Hall of Mirrors and various swings and puzzles. The museum has been very well received by the public and, according to its own information, receives an average of 150,000 visitors per year. It can thus finance itself almost exclusively from its own income; at least that was the situation before the COVID-19 crisis.

The Mathematics Adventure Land ("Erlebnisland Mathematik") in Dresden, see Chapter 7 of this handbook, is also devoted entirely to mathematics and offers a wide range of exhibits on 1,000 m² of exhibition space; around 100 participatory exhibits, many of which can be understood intuitively. For many of them, meticulously prepared background information is available. Various educational tools and experiences are made for schools and there are numerous curricula references for grades 5 to 10.[ae]

Following these major mathematics exhibitions in Germany, other places of science will now be discussed. With numerous institutions dedicated to science communication in the broadest sense, communication on the topic of science has been institutionalized. The participatory and dialogic character of science communication has been expanded with them and partly perpetuated.

Houses of Science

The success of science communication in Germany in the early 2000s, as outlined above, also gave some cities the opportunity to raise their profile as "science cities" and expand their location marketing thematically. So-called Houses of Science ("Haus der Wissenschaft") were created and supported by local initiatives, for example the House of Science in Bremen (2005) and the House of Science in Braunschweig (2007). These "houses", which served the citizens' dialogue, helped the cities to distinguish themselves as science locations. For example, topic- and location-related small

---

[ae] http://www.erlebnisland-mathematik.de/en/.

exhibitions, discussion formats on selected topics, as well as lectures by researchers followed by discussions were offered to the general public. The classic lecture format was broken up or supplemented by a thematic discussion with the audience. The House of Science in Bremen, for example, describes its mission as follows, "We offer visitors from Bremen and the surrounding area a broad program: with exhibitions, lectures and discussions, we bring them closer to current topics from science and research in an understandable way (...)". One example: lectures by scientists on the unique fauna of the Wadden Sea are combined with discussions on environmental issues, such as with regard to the competing operation of the economically important Bremen port in neighboring Bremerhaven.[af]

Furthermore, the House of Science acts as a hub for local research institutions by providing insights into the work of universities, colleges, and research institutes in Bremen and also specifically informing school classes about the latest results from the "world of research" of these institutions. Dialogue with teachers and representatives from the industry is also sought. For this purpose, special dialogic formats have been created. In Bremen for example, the Forum Science and School ("Forum Wissenschaft und Schule") that takes place five times a year, and the so-called Science Matinee ("Wissenschaftsmatinee"), every Saturday at 11:00 a.m.

Examples of what is offered by the House of Science in Braunschweig, where a large technical university is located, include the Digital Marketplace ("Digitaler Marktplatz") with topics and ideas for digital transformation, where interested parties can choose between 25 experts from science and business for a one-on-one discussion, which then takes place, similar to speed dating, at individual tables in a large hall in the city. Also, the website's description of the temporary Make Your School project in Braunschweig, which ties in with the idea of the School Labs (i.e., "learning by doing" at an extracurricular learning venue), reads:

"In the Make Your School project, students can help shape their school and try their hand at programming, tinkering and crafting. The so-called Hack-days are intended to help improve digital education at schools

---

[af] www.hausderwissenschaft.de/.

and introduce young people to the range of digital and electronic tools. In addition, the aim is to promote the ability to identify problems and opportunities for improvement and to develop solutions independently. WiD wants to use the project, which is significantly supported by the Klaus Tschira Foundation, to provide new impetus for everyday school life."[ag]

Temples of Science

Thematic or local features of on-site research were addressed additionally. With lots of effort impressive centers were established to worship science. That's why we call them temples of science. We give three prominent examples below:

Climate house

The Hanseatic and port city of Bremen with the location of the Alfred Wegener Institute, Helmholtz Center for Polar and Marine Research (AWI), in neighboring Bremerhaven founded the so-called Climate House ("Klimahaus"), a large science center with expressive architecture in the form of a boat, whose 18,800 m² area is dedicated to the topics of weather, climate, climate change, and climate zones of the earth.[ah]

In the Climate House, visitors can stroll through the different climate zones of the earth, with some parts having spectacular staging and effects so we are in the realm of edutainment here. Mathematics and its contribution to the above-mentioned range of topics is only mentioned in passing. The House has welcomed over four million guests since it opened in 2009. According to its own figures, more than 457,000 visitors came in 2019 alone.

Phaeno

A weighty temple of science is the Science Center phaeno in Wolfsburg, which opened at the end of 2005 after 4 years of construction and building costs of 90 million euros. Its guiding themes are "Life, Vision, Energy, Dynamics, Sense and Math". The access to the topics is intuitively playful and emotional: "This is how you experience surprising moments that astonish you and puzzle you. This sudden emotional reaction awakens the spirit of discovery and paves the way for further engagement with the phenomenon", phaeno states on its website. This means a psychological understanding of the access to knowledge: Not only is the play instinct to be activated, but emotions are to be aroused in a targeted manner. This shows how sophisticated the communication process has become in science centers.[ai]

Arousing emotions and people's curiosity is done through "an adventure landscape for extraordinary phenomena," as it says on the phaeno website. This claim manifests itself in a futuristic building that the star architect Zaha Hadid had artfully sculpted from concrete and serves as a counterweight to the "car city" of Volkswagen AG located opposite of it.[aj] The phaeno Foundation is the sponsor and the main benefactor is the city of Wolfsburg. Additional funds are generated by the Friends of the phaeno Wolfsburg ("Freundeskreis phaeno Wolfsburg e.V.").

Inside the sleek building are several hundred experiment stations and also works of art (about ten percent of the exhibits). "From fire tornadoes to water whirlpools, phaeno offers countless opportunities to let curiosity run free," it says under experiences for kindergartens, schools, families, youth groups, and seniors, which are partly supplemented by activities to do at home. Among these are materials on mathematics, for example a task sheet on the "magic of shapes and patterns" for grades 4–6. Phaeno's math show called "Math Makes You Happy" is limited to a few striking examples such as the geometry of the soccer ball, fractals, the Möbius strip, folding paper several times to illustrate exponential growth, and number tricks; a good example of math edutainment which can also be seen on YouTube.[ak]

Futurium

More sober but no less modest is the large, black, glittering structure of the Futurium in Berlin, whose shareholders are predominantly research-related, consisting of the federal government (BMBF), the above-mentioned science organizations (WiD's shareholders), and various foundations and companies.[al] The Futurium was founded in 2014 to serve as a "forum for presentation and the facilitation of dialogue on scientific, technical

---

[ai] www.phaeno.de/en/.

[aj] https://www.phaeno.de/virtueller-rundgang/.

[ak] https://tinyurl.com/uktkxp3p.

[al] https://futurium.de/.

and social developments of national and international significance, and to conduct a scientific-based social discussion on shaping the future. Futurium constitutes an independent platform for hosting a dialogue and establishing links between the state, science, business and society," which is how the Futurium describes itself on its website. Its three pillars are called discover, trial and error, and discussion. In fact, there seems to be a certain return to the shareholders' claim as education and public dialogue is emphasized here. Mathematics is thematized in the Futurium according to Galileo Galilei's witty remark that the book of nature is written in the language of mathematics with rather common exhibits such as the expression of the Fibonacci sequence in nature (shell casing of the nautilus, arrangement of sunflower seeds, etc.), self-similar cabbage plants and the golden ratio and golden angle (visualized by means of a spirograph).

Major events

Parallel to the various permanent exhibitions, major regional events on the theme of science developed, for example the "Explore Science — The Natural Science Experience Days" ("Die naturwissenschaftlichen Erlebnistage",[am] Figure 9) of the Klaus Tschira Foundation in Mannheim and Bremen. "Since 2006, all inquisitive minds can explore, experience and discover for themselves. The offers for all age groups range from interactive exhibitions, numerous hands-on activities, workshops and stage shows to competitions and experimental lectures", says the website of the annual festival. It continues: "With Explore Science, the Klaus Tschira Foundation wants to awaken young people's interest in scientific topics. The central concern here is that children and young people are not 'served' answers but are given the opportunity to discover scientific phenomena for themselves." The entertainment value of Explore Science is huge with elaborate stage shows.

Participative formats

Now some words on the trends of today (2022). During the past 20 years, the trend has been toward participatory formats, i.e., visitors from different

---

[am] www.explore-science.info.

Figure 9: The target groups became younger over the years: children of preschool age at the Explore Science Festival in Mannheim. Copyright Klaus Tschira Foundation.

target groups should not simply carry out predefined experiments, but become active themselves and contribute their own ideas. In this sense, the new format of Makerspaces has emerged in the last 10 years, where a certain amount of technical equipment and know-how is provided and citizens can realize their own predominantly technical wishes and ideas. Relevant ideas and initiatives have been successful also because they have been able to rely on existing spaces, for example the Houses of Science.

Raspberry Pi, the do-it-yourself construction of mini-computers, experienced a renaissance and Repair Cafés combine enthusiasm for technology and environmental awareness.

In 2002, the Fraunhofer Institute IAIS launched the Roberta initiative to introduce young people, especially girls, to programming. With the help of small kits and the programming language NEPO, small robots can be put into action with guidance (e.g., via "open Roberta Labs"). The educational initiative was initially funded by the BMBF (2002–2006), then by the European Union (2005–2008) and continues to exist as "open Roberta".[an] Even if the examples above are sometimes more about manual

---

[an] open-roberta.org.

dexterity and sometimes more about programming skills, there is almost always a reference to mathematics and the formats captivate by activating the creativity of young people.

So-called Real Labs ("Reallabore") have been invented to test, in a small and defined framework, a certain innovative technology product or service under real circumstances and in close feedback of scientists of different disciplines or citizens. Topics of interest can be sustainability, transformation, integration, innovative development, etc. When citizens are involved it may be called a citizen science project.[ao]

Citizen Science

Citizen science projects, in which citizen participation is used for the benefit of science, have also become quite trendy in recent years. There are some very impressive examples from biology, ecology, environmental protection and related subjects, such as the Nature and Biodiversity Conservation Union in Germany (NABU) calling on citizens to count insects, birds, etc. in diversity monitoring projects.[ap]

In the yoyo@home project for distributed computing, anyone can make the computing power of their home computer or smartphone available to science.[aq] In this context, reference should be made to an ambitious international project on the search for prime numbers, the Great Internet Mersenne Prime Search (GIMPS), a collaborative project on the computer-aided search for Mersenne prime numbers. Since 1996, hundreds of thousands of volunteer users worldwide have used their PCs to search for Mersenne prime numbers of the form $2^p - 1$. As of October 2021, 47 Mersenne numbers have been found and confirmed.[ar]

A rather playful mathematics project with a strong participatory character is polytopia.eu: interested people are invited to adopt, name

---

[ao] w www.buergerschaffenwissen.de/en.

[ap] https://en.nabu.de/get-involved/index.html.

[aq] https://www.rechenkraft.net/yoyo/.

[ar] https://www.mersenne.org/.

and craft polyhedra. In this way, thousands of polyhedra are lifted out of abstraction and brought into realization.[as]

Little Scientists' House

In conclusion, it can be said that there is now (2022) a certain return to the initial initiatives for the extracurricular teaching and learning of natural sciences. This trend manifests itself, for example, in the initiatives of the nonprofit foundation Little Scientist' House, which is, in its own words, "Germany's largest childhood education initiative for daycare centers, after-school programs and elementary schools." This also shows that, much like the trend to target a young audience mentioned previously in connection with the Math Advent Calendar and the "Explore Science" the trend is to continuously address even younger target groups. The Little Scientists' House is aimed at children of kindergarten age. In line with the UN's 17 Sustainable Development Goals (SDGs), the focus here is explicitly on educational opportunities and sustainability aspects.[at]

In order to interest and educate young children in the natural sciences, the Little Scientists' House has created a nationwide opportunity to enable educators to teach natural sciences "at a uniform level" (see webpages). Under the heading "Mehr Bildungschancen für alle" (More Educational Opportunities for All), the foundation's website states: "The non-profit Foundation Little Scientists' House is committed to good early education in the fields of mathematics, computer science, natural sciences and technology ("MINT") throughout Germany — with the aim of making girls and boys strong for the future and empowering them to act sustainably."

This approach brings us back to the beginning of this chapter, where we started with formats of out-of-school teaching and learning (School Labs etc.). It is also interesting to note that the Little Scientists' House is a nationwide initiative for the further training of educators that is funded by several private foundations in addition to the BMBF. Finally, it is worth mentioning in this context that 10 years ago the Deutsche Telekom

---

[as] https://www.polytopia.eu/en/.

[at] https://www.haus-der-kleinen-forscher.de/en/about-us.

Foundation initiated and funded the establishment of the "German Center for Teacher Training" ("DZLM")[au] to improve teacher training. In the meantime, the DZLM has become permanent in the form of a department of the Leibniz Institute for Science and Mathematics Education (IPN). Little Scientists' House and DZLM are an expression of the realization that teacher training in Germany must also be improved if students from Germany are to perform better in internationally comparative tests. The Science on Stage initiative, a nonprofit initiative that has been bringing together STEM teachers with "outstanding teaching ideas" for 20 years now, is also committed in this sense, as it says on its website.[av]

Outlook

For the future, it is foreseeable that students' knowledge of mathematics will only improve if more financial resources are invested in the German school system. Better supervision ratios (more teachers) would be important here. Incidentally, this would also allow more teachers to participate in further training which would result in better teachers. In addition, a better harmonization of learning content in Germany across the federal states is desirable, as well as binding minimum standards in mathematics instruction and for the Abitur mathematics exam, especially with regard to subject content. However, this is a discussion that must be conducted in a political context and together with mathematics didactics. In addition, the task of promoting digitization in schools is still unresolved in Germany in 2022, bearing lots of potential for teaching mathematics.

In the area of extracurricular learning, there will be a continuation of successful formats, such as the School Labs or similar. Edutainment will also remain in the STEM subjects, whether being a part of science fairs, science festivals, or science centers. The need for dialogic formats will also remain and continue to grow; not least of all, this is due to the challenges ahead, such as the energy transition and tackling the climate crisis. Here, solutions will only be achieved through dialogue and consensus with the population.

[au] https://dzlm.de/en/international-visitors.

[av] www.science-on-stage.de.

Examples include the expansion of renewable energies, especially wind energy, and geoengineering processes, such as technologies for CO_2 capture and storage. In this context, the voice of science will play an important role as a reliable corrective to political measures, similar to what we experienced in the course of the COVID-19 crisis in Germany 2020/2021.

Biographical Note

Thomas Vogt has studied geology, German literature, and science journalism in Berlin. He wrote freelance articles for the Berlin daily newspaper *Der Tagesspiegel* (1998), then was employed as a science journalist by Helmholtz Center Jülich (1999–2002), the German Aerospace Center (DLR) and the Leibniz Association, where he was a spokesperson as well (2006–2007). In his various positions, he contributed each year to the Science Years in Germany from 2000 (year of physics) to 2006 (year of computer science). In the Year of Mathematics (2008), he became a senior writer and editor on behalf of Deutsche Mathematiker-Vereinigung at the MCBO at Prof. Günter Ziegler's chair. Today he is the spokesperson and head of the Math Media Office of the German Mathematical Society at Freie Universität Berlin,[aw,ax] www.mathematik.de, Contact via vogt@mathematik.de.

References

[1] Peters, H. P., Lehmkuhl, M., and Fähnrich, B. "Germany: Continuity and change marked by a turbulent history". In Gascoigne T., Schiele B., Leach J. *et al* (eds.). *Communicating Science: A Global Perspective*. ANU Press. https://press-files.anu.edu.au/downloads/press/n6484/html/ch14.xhtml.

[2] Baumert *et al.* (2001). *PISA 2000: Basiskompetenzen von Schülerinnen und Schülern im internationalen Vergleich* (1st edn.). Springer, 548 pp.

[3] Behrends, E., Crato, N., and Rodrigues, J. F. (2012). *Raising Public Awareness of Mathematics* (1st edn.). Springer, 404 pp.

[aw] https://www.mathematik.de/kontakt-dmv.

[ax] https://madipedia.de/wiki/Thomas_Vogt.

[4] Picker, S. H. and Barry, J. S. (2000). "Investigating pupils' images of mathematicians". *Educational Studies in Mathematics*, 43(1), 65–94.

[5] Fink, T. and Mao, Y. (1999). *The 85 Ways to Tie a Tie: The Science and Aesthetics of Tie Knots*. Fourth Estate, 144 pp.

[6] Behrends, E., Gritzmann, P., and Ziegler, G. M. (eds.). (2016). *PI & Co.: Kaleidoskop Der Mathematik* (2nd edn.). Springer, 421 pp.

Epilogue

National Museum of Mathematics (MoMath): A Place for All People

As I write this epilogue, it is September 13, 2021, exactly 18 months to the day since MoMath — and much of New York City — shut down in response to the COVID-19 global pandemic. But MoMath's pandemic story started many months earlier, back in January of 2020, when a young astrophysicist from Harvard came to visit and arrived wearing a mask. She had heard that a novel virus was sickening people in China and assumed that if it were spreading in China, it would also be spreading in NYC. I couldn't help but hear her words continuing to echo in my head in the days that followed.

By the time the Museum, the city, and the country largely locked down, Museum managers had already been planning for the better part of a month. We had settled on Zoom as our preferred mode of interaction with patrons and had a good understanding of how it could work, including what it would cost, how many virtual rooms we would need, etc. But still, none of us expected to use those plans: like school administrators who dutifully conduct fire drills year after year while never anticipating an actual fire, we felt good that we were prepared for an emergency we never expected to happen.

But happen it did. Everything changed on March 13, 2020, when much of New York City locked down and MoMath closed its doors for what would turn out to be well over a year. By then, however, MoMath's foray into the world of online programming had already begun: on March 12, just one day before the shutdown — the very same day that MoMath welcomed what would turn out to be its last visitors for 15 months — we also welcomed our first online school visitors. Thanks to the prescient words of that bright young astrophysicist, MoMath was able to continue its programming without missing a beat.

But none of that is to say the pivot was easy. Turning from a hands-on museum into a provider of online programming overnight placed us on a steep learning curve. In addition to mastering the vagaries of Zoom, we also had to figure out how to take programs that focused on live, in-person interactions, often using physical materials, and turn them into online events that would work for participants who were sheltering at home. We had to carefully consider the key components of MoMath's successful outreach, and determine how to translate those characteristics into an online forum. The physical exhibits had been critical to our success — it's easy to draw someone in with a wall of shimmering laser light, a series of giant colorful knobs to turn, or an odd-looking tricycle with square wheels on which to ride. Without a physical experience as a base, we needed to revisit some basic questions. Why do some people have an uncomfortable relationship with mathematics, how did MoMath work to change that, and how could those strategies be reimagined to work in an online world?

For many people, their first experience of mathematics comes in a classroom, with a focus on rote memorization to reach a predetermined answer. For some, this comes naturally — like a child born with perfect pitch, these students are identified as exceptional. But that's where the analogy ends. Children who are not initially at ease in the world of music are encouraged with adages like "practice makes perfect". We expect that anyone, with enough effort, can become a reasonably proficient musician. How wonderful if this same logic were routinely applied to mathematics: with effort and encouragement, development of basic mathematical skills should be broadly attainable for all.

Recognizing this goal during the initial creation of MoMath, the design team deliberately stayed away from making the Museum look like "math" — there are few numbers, no chalkboards or calculators, and the exhibits are more likely to be thematically related to the arts, architecture, and culture than they are to the familiar symbols one sees in a math textbook. Our goal was to warmly welcome people from the minute they walked in…and to avoid making them feel like they had entered a place in which they didn't belong. How could we replicate that in online programs? The key for us was in the broad range of programming we could now provide.

From origami artists to puzzle designers, from popular authors to klezmer musicians, and from sports analysts to professional dancers... our audience was enticed by programming that beckoned and then, only after they entered, gently revealed a connection to mathematics. This parallels the experience many have in the physical Museum...enticed by exhibits that draw them in, many report having a newfound appreciation of mathematics by the time they depart.

At one popular program, we paired Julie Gerberding, an infectious disease expert who was the first woman to serve as the director of the U.S. Centers for Disease Control and Prevention (the CDC) and who later joined Merck as president of vaccines, with Eric Schmidt, former CEO of Google and current Chair of New York State's 16-member Blue Ribbon Commission developed in the wake of the pandemic. More than 250 people participated in this highly engaging and informative discussion about the current state of the global health crisis, the progress of vaccine development, and the expected long-term impact of the pandemic on health care, business practices, and psycho-social norms and expectations.

Another popular series of programs featured Dutch artist Anton Bakker, whose physical art exhibition (initially planned to open in spring 2020) was converted into a fully accessible online art show that allowed visitors not just to move through the gallery and view the art, but to interact with the sculptures from many different perspectives. More than 1,000 people from all over the world attended 14 programs presented by Anton over the course of 14 months.

Breathtaking photographs of birds were a central feature of a popular program — attended by almost 600 attendees — featuring Peter Cavanagh, an Emeritus Professor of Orthopedics and Sports Medicine from the University of Washington. Robert Leonard, Professor of Comparative Literature, Languages, and Linguistics at Hofstra University on Long Island captivated an audience of more than 400 people "whodunit style", by showing how pattern recognition, one of the basic building blocks of mathematics, could also be applied to language and sentence structure to solve actual crime mysteries. And for the first time ever, the uniquely challenging MoMath Masters adult math tournament was open to participants from around the world, with more than 400 people participating and the ultimate winner — a first-time participant — joining remotely from northern California.

Online programming also allowed MoMath to expand its reach– no longer was its audience limited to those who could make their way into Manhattan. During the pandemic, for example, one young brother and sister from India routinely joined MoMath programs from the darkness of their bedroom while they were supposed to be asleep; a young student who was relocated from New York to Russia during the shutdown continued to join MoMath programs from afar; and several children in California arose at 6:00 am daily for the chance to participate in MoMath's online summer camp. On several occasions, people even participated in MoMath programs from moving vehicles! The online auditorium became a welcoming space for people from all over the world — in more than 100 countries — to come together for shared experiences of mathematics.

In addition, top-flight presenters could suddenly be on a world stage without the need for travel, time away from home and work obligations, and extensive funding. Now, with the flip of a switch and the push of a button, and at the cost of just an hour or two of time, those who had something they wished to share found a global audience of enthusiastic people eager to hear from them, while those who might never have come into contact with renowned mathematicians suddenly had the opportunity, no matter where they were, to engage with some of the top minds in the field. Fields Medalist Manjul Bhargava delighted audience members with a card trick that has deep mathematical underpinnings; YouTube phenom Grant Sanderson (3Blue1Brown) shared his life story with eager audience members; and German mathematician and computer scientist Jurgen Richter-Gebert delighted participants by using a bit of math to "read their minds".

This ability to bring top presenters together had another important benefit, too. When we wanted to show young women and Black and brown youth that they too belong in the world of math, we brought together small groups of highly accomplished professionals in mathematics — who happened to be women, or Black, or LatinX — to make that case emphatically. These sorts of programs would have been difficult — if not impossible — to orchestrate if we needed to coordinate schedules and travel logistics for so many people at the same time. All told, almost 1,400 people attended four programs highlighting 21 successful mathematicians and engineers, all members of groups underrepresented in mathematics.

While finding topnotch presenters and bringing them to the MoMath "stage" was a critical prerequisite to online programming, simply identifying engaging presenters was not sufficient — we also had to think carefully about the mode of online presentation. In the pre-pandemic world, live events were compelling not just because of the presentation itself, but because of the social opportunities afforded to participants. Audience members mingled upon arrival, chatting casually with each other, with Museum personnel, and even with the presenter. And after a talk ended, people turned to their seatmates to share their thoughts, or they headed to the stage to chat — or grab a quick selfie — with the presenter. We realized that we needed to mimic this in our Zoom programs, so we allowed audience interaction right from the start, with participants sharing their location and making connections with their fellow participants. As programs concluded, we encouraged participants to linger, unmuting to share an audible reaction and allowing time for more personal discourse with each other and the presenter.

We also found an unexpected silver lining from virtual programming. The cloak of anonymity one can find in the online world, where participants are typically only identified by a first name or even a pseudonym, can actually encourage discourse among those who might be hesitant to ask questions in person. How wonderful it is to see questions asked...and sometimes answered...by members of the audience.

On the other hand, the world of online presentations posed some unique challenges as well. Programs that relied on tangible materials had to be reimagined: for example, physical code wheels were turned into online interactives and a group mosaic construction was turned into a virtual build, with participants sending photos that were electronically combined to create the final artwork. In-person gatherings had to be recast as well: a math festival that typically took place on a crowded public plaza turned into a mathematical vaudeville show attracting more than 1,200 participants, with a steady stream of online presenters challenged to convey an interesting and engaging bit of mathematics in 15 minutes or less.

Connectivity was not universal and reaching disadvantaged populations was difficult. Additionally, after more than a year of remote interaction, Zoom fatigue was starting to set in.

When hope finally arrived in the form of a number of long-awaited vaccines, the Museum reopened, live presentations resumed, and it was back to business as usual.

Or was it? It turns out this epilogue itself is not the end of the story, but rather a new beginning. As I write this, we have begun to ascend yet another new (and steep) learning curve. While the Museum has reopened its physical space and visitors have begun to return, our newfound global audience remains. Having enjoyed MoMath's engaging online experiences, many are eager for the programs to continue. And so, we move forward, running programs live in the Museum while simultaneously trying to include remote participants from around the world. The key seems to be in connecting the audiences; in breaking down the virtual walls so that the two groups become one and a single experience is shared by all. The pandemic has underscored the fact that math is indeed a universal language and a global, intergenerational connector, transcending language and culture. It allows disparate people from around the world to build new human connections, uniting people over a shared experience, and it helps us find universal truths that can be both beautiful and inspiring. While it took a pandemic to force us to embrace online programming, there is now no going back. We've thrown open the doors to this new world and, as we gradually return to in-person events, gatherings, and visits, we intend to keep the doors of mathematical inquiry and engagement propped wide open, welcoming all into the wonderful world of mathematics.

Cindy Lawrence
Executive Director and CEO
National Museum of Mathematics
New York, USA

Index

CPSIA information can be obtained
at www.ICGtesting.com
Printed in the USA
JSHW041437130123
36111JS00002B/6

9 789811 253065